MAKING SPATIAL DECISIONS

USING GIS
AND LIDAR
A WORKBOOK

Kathryn Keranen ▪ **Robert Kolvoord**

Esri Press Academic
REDLANDS | CALIFORNIA

To Sammy—unfailing friend, confidante, and tireless listener.

—KK

To my teachers, for the motivation and inspiration they provided.

—RK

CONTENTS

Preface _____ ix

Acknowledgments _____ x

Introduction _____ xi

MODULES 1 THROUGH 5 SCENARIO _____ 1

Module 1: Basic lidar techniques _____ 2

Project 1: Baltimore, Maryland _____ 4

Project 2: San Francisco, California _____ 20

Project 3: On your own _____ 27

Module 2: 2D and 3D models (raster and terrain) _____ 30

Project 1: Baltimore, Maryland _____ 33

Project 2: San Francisco, California _____ 46

Project 3: On your own _____ 53

Module 3: Volumetric analysis and shadow maps _____ 56

Project 1: Baltimore, Maryland _____ 59

Project 2: San Francisco, California _____ 68

Project 3: On your own _____ 72

Module 4: Visibility analysis comparison _____ 74

Project 1: Baltimore, Maryland _____ 77

Project 2: San Francisco, California _____ 86

Project 3: On your own _____ 89

Module 5: Surging seas _____ 92

Project 1: Baltimore, Maryland _____ 94

Project 2: San Francisco, California _____ 99

Project 3: On your own _____ 103

MODULES 6 THROUGH 7 SCENARIO _____ 105

Module 6: Corrected 3D campus modeling _____ 106

Project 1: James Madison University, Harrisonburg, Virginia _____ 108
Project 2: University of San Francisco, San Francisco, California _____ 113
Project 3: On your own _____ 116

Module 7: Location of solar panels _____ 118

Project 1: James Madison University, Harrisonburg, Virginia _____ 120
Project 2: San Francisco University, San Francisco, California _____ 125
Project 3: On your own _____ 129

Module 8: Shoreline change after Hurricane Sandy _____ 132

Project 1: Deal, New Jersey _____ 135
Project 2: Pelham Manor, New York _____ 142
Project 3: On your own _____ 146

Module 9: Forest vegetation height _____ 148

Project 1: George Washington National Forest, Virginia _____ 151
Project 2: Michaux State Forest, Pennsylvania _____ 160
Project 3: On your own _____ 165

Module 10: Depressional wetland delineation from lidar _____ 168

Project 1: Wakulla, Florida _____ 171
Project 2: Pasco County, Florida _____ 181
Project 3: On your own _____ 185

Image and data credits _____ 187
Data license agreement _____ 197

PREFACE

We wrote this book to help you combine an exciting new data source—lidar—with geospatial analysis tools to help government, business, nongovernmental organizations, and other entities that rely on geographic information system (GIS) technology to make critical decisions.

Making Spatial Decisions Using GIS and Lidar: A Workbook opens the world of lidar data for you to analyze, interpret, and apply to various scenarios. You will make the types of decisions that affect an agency, a community, or a nation. The projects in this book use lidar data as part of the GIS analysis and mapmaking process. The projects are focused on problem solving, and will help you hone your critical-thinking skills. We have chosen scenarios that are relevant, challenging, and applicable to a broad range of disciplines, not just geography. We also think you will enjoy finding solutions to interesting problems.

Making Spatial Decisions Using GIS and Lidar is a college-level textbook that presumes you have some GIS experience. Technical prerequisites are outlined in the introduction, which also explains how to use the book, organize your workflow, and evaluate your work products. The introduction also includes a summary of all 10 modules.

Making Spatial Decisions Using GIS and Lidar comes with GIS data, worksheets, and other documents you will need to complete the projects. You can download these materials and access a 180-day trial of ArcGIS from the Esri Press "Book Resources" web page: esripress.esri.com /bookresources. Click the *Making Spatial Decisions Using GIS and Lidar* book title, and then click the appropriate links under "Resources" to access the trial software and download the exercise data, worksheets, and documents. Information about online resources is provided in the book.

The worksheets provided on the Esri Press "Book Resources" web page will help you follow the activities in the book. The worksheets provide a place where you can answer questions presented in the book and keep track of the work you have completed. Your answers will help you analyze and answer the larger problems in each module. Do not lose sight of the forest (the overarching problem) by spending too much time looking at the trees (the individual questions). Your instructor has supplemental resources available to assist you with the projects in this book.

We have been excited about the possibilities that lidar data offers for a number of years, and these modules give us a chance to share the power of combining lidar and GIS with you. This book stems from our many decades of experience working with students and teachers doing spatial analysis and problem solving. We hope you enjoy using this book as much as we enjoyed writing it!

Kathryn Keranen and Robert Kolvoord

ACKNOWLEDGMENTS

We would like to thank Mr. Jean Lorber, Land Protection Specialist at The Nature Conservancy Virginia Field Office in Charlottesville, Virginia, for his advice and support on the Forest Vegetation Height module.

We would like to thank James Madison University, Florida International University, and Stuart Hamilton (http://www.virginialidar.com) for sharing data with us.

We would like to thank Mr. Bill Ryan and his students at Colonial Heights High School in Colonial Heights, Virginia, for their efforts in testing the modules in the book and providing feedback.

We would also like to thank the staff at Esri Press for all their assistance in bringing this book from concept to print.

Finally, we want to acknowledge all of the students and teachers in the Geospatial Semester with whom we get to work and who contributed to this project.

INTRODUCTION

Remote sensing data and imagery often come from satellites orbiting high above the earth. These satellites use "passive" remote sensing, merely collecting light that is reflected or refracted from objects on the earth's surface or in the atmosphere. But what if instead of relying on the sun, you could actively illuminate objects on the surface of the earth? This is the genesis of lidar (a combination of light and radar), a 50-year old technology that uses lasers to provide very high-resolution imagery of various objects. Originally used to map clouds in the atmosphere and to provide high-resolution altimetry of the moon for Apollo missions, lidar has now become a commonly used technology to generate high-resolution maps.

This book is the third in the Making Spatial Decisions series and continues the focus on scenario-based problem solving using an integrated workflow. The scenarios here feature the use of lidar as the primary data source and lidar-related analysis as the primary skill focus. Lidar is a powerful, and increasingly popular, data source used to create elevation and terrain models with high accuracy.

The workflow process used in the book is as follows:

1.   Define the problem or scenario.
2.   Identify the deliverables needed to support decisions.
3.   Document, set environments, and examine the data.
4.   Perform analysis starting with a basemap.

Each of the 10 modules in this book follows the format used in *Making Spatial Decisions Using GIS: A Workbook* and *Making Spatial Decisions Using GIS and Remote Sensing: A Workbook*.

Project 1 gives step-by-step instructions to explore a scenario. Users answer questions and prepare maps to include in a presentation of their analysis. Users decide how to resolve the central problem in the given scenario.

Project 2 provides a slightly different scenario and the requisite data without step-by-step directions. Users must apply what was learned in project 1.

Finally, each module includes an "On your own" section that suggests different scenarios users can explore by locating and downloading data and reproducing the analysis from the guided projects.

In these modules, we presume that you have used ArcGIS software before and that you can do the following basic tasks:

- Navigate and find data on local drives and network drives.
- Name files and save them to a known location.
- Use ArcCatalog to connect to a folder.
- Use ArcCatalog to preview a data layer and look at its metadata.
- Add data to ArcMap by dragging layers from ArcCatalog or using the Add Data button.
- Rearrange layers in the ArcMap table of contents.
- Identify the Table of Contents and the Map windows in ArcMap and know the purpose of each.
- Use the following tools:
  - Identify
  - Zoom in
  - Zoom out
  - Full extent
  - Pan
  - Measure
- Symbolize a layer by category or quantity.
- Open the attribute table for a data layer.
- Select features by attribute.
- Label features.
- Select features on a map and clear a selection.
- Work with tables and edit attribute values.
- Make a basic layout with map elements.

Topical instructions are given in the book. If more detailed instructions are needed, ArcGIS for Desktop provides the following options:

- Use ArcGIS Help to ask a question or look up a keyword, such as a "tab," "menu option," or "function." If you are online, it is better to use the web-based Help option by accessing the ArcGIS Resources website, http://resources.arcgis.com, for an up-to-date version of ArcGIS Help included with the software.
- The various geoprocessing tools can be accessed by using ArcToolbox or through the Search For Tools option in the Geoprocessing menu. When you search for a tool, an explanation and a link to the tool appears.

Making Spatial Decisions Using GIS and Lidar was written specifically for ArcGIS 10.2 for Desktop and was tested in version 10.3 of the software. The projects use the LAS Dataset toolbar, which toggles between lidar points and TIN-based surfaces (TIN stands for triangulated irregular network). The LAS Dataset toolbar allows you to view lidar attribute data. The lidar datasets used in this book have all been downloaded from public domain websites. Each website is credited, and metadata are provided for each of the datasets.

MAKING
SPATIAL
DECISIONS
USING GIS
AND LIDAR

INTRODUCTION

MODULES 1 THROUGH 5 SCENARIO

Modules 1 through 5 focus on a series of problems to be solved by a GIS Company for two different urban locations: Baltimore, Maryland, and San Francisco, California. The city governments of Baltimore and San Francisco are aware of the new visualization and analysis capabilities that the use of lidar will bring to their municipalities. Your company has been contracted to do the following:

1. Investigate the integrity of the lidar data the cities have obtained.
2. Create highly detailed 2D and 3D city models.
3. Perform a volumetric analysis and produce shadow maps and visibility/viewshed analysis for cellular network planning.

Your company has studied the requirements of the two cities and, upon closer examination, decided to divide the project into five smaller parts (modules 1 through 5):

- Module 1: Basic lidar techniques

- Module 2: 2D and 3D models (raster and terrains)

- **Module 3: Volumetric analysis and shadow maps**

- **Module 4: Visibility analysis comparison**

- Module 5: Surging seas

Note: Modules 1 through 5 all use the same data; therefore, only one data and one results folder exist for each project in the exercise data: baltimore_data and baltimore_results, and sanfrancisco_data and sanfrancisco_results.

MODULE 1
BASIC LIDAR TECHNIQUES

SCENARIO

One of the first tasks a geographic information systems (GIS) department using lidar data should perform is to check the quality of the data delivered by the data provider. The department can assess the data quality by looking at the points delivered by the data provider to determine the coverage and density of the data. Pertinent information about the LAS data is found in both the metadata for the lidar flight and the statistics generated using the LAS Dataset Properties tool. After the initial information is gathered, the GIS analyst must measure the distance between points and look for noise or other uncalculated points. After these preliminary investigations, the data can be viewed as a point cloud and a surface in two dimensions (2D) and three dimensions (3D).

Projects included in this module

- **Project 1:** Baltimore, Maryland
- **Project 2:** San Francisco, California
- **Project 3:** On your own

Student worksheets

Project 1: Baltimore, Maryland

- File name: 1a_basics_worksheet.docx
- Location: EsriPress\MSDLidar\student\1_modules1_5\1basics\project1_documents

Project 2: San Francisco, California

- File name: 1b_basics_worksheet.docx
- Location: EsriPress\MSDLidar\student\1_modules1_5\1basics\project2_documents

Objectives

- Explore the properties and statistics of a LAS dataset.
- Create a LAS dataset with appropriate coordinate systems and point spacing.
- Measure heights and distances between points.
- Visualize a LAS dataset as a TIN (triangulated irregular network).
- Conduct interactive surface analysis.
  - Classes
  - Returns
    - Elevation
    - Slope
    - Aspect
    - Contour
- Edit and reclassify noise points.
- Visualize a LAS dataset as a 2D and 3D surface.
- Visualize a LAS dataset in a 3D profile.

Baltimore, Maryland

Recommended deliverables

- **Deliverable 1:** A LAS dataset statistics file.
- **Deliverable 2:** A basemap showing the location of the LAS dataset.
- **Deliverable 3:** A map with three data frames showing points symbolized by class, an elevation TIN, and a TIN with a ground filter. A brief written analysis should be included on the map.
- **Deliverable 4:** A document containing a profile, a 3D view, and a written description.

The questions asked in this project are both quantitative and qualitative. They identify key points that should be addressed in your analysis and final presentation.

Document your work, set environments, and examine the data

1. Open ArcMap. (For these exercises, the Getting Started window is not needed. Select the "Do not show this dialog in the future" box at the bottom of the window so it does not appear in the future).

You need to add descriptive properties to every map document you produce. You can use the same descriptive properties for every map document in the project or customize the documentation from map to map.

2. From the File menu, select Map Document Properties.

The Map Document Properties dialog box allows you to add a title, summary, description, author, credits, tags, and hyperlink base.

DATA (Current Workspace)   \student\1_modules1_5\baltimore_data
RESULTS (Scratch Workspace)   \student\1_modules1_5\baltimore_results

MAKING
SPATIAL
DECISIONS
USING GIS
AND LIDAR

1

BASIC LIDAR
TECHNIQUES

3. Enter the map document properties for this project as shown in the following figure. Next to Pathnames, select the "Store relative pathnames to data sources" check box, which tells ArcMap to store relative pathnames to all of your data sources. Click OK.

Storing relative pathnames allows ArcMap to automatically find all relevant data if you move the project folder to a new location.

One of the most important things to do in a geospatial project is carefully keep track of the data and your calculations. You will work with a number of different files, and it is important to keep them organized so you can easily find them. The best way to do this is to have a folder for your project that contains a data folder. For this project, **baltimore_data** will be your project folder (EsriPress\MSDLidar\student\1_modules1_5\baltimore_data). Make sure that it is stored in a place where you have write access. You can store your data inside the **baltimore_results** folder (EsriPress\MSDLidar\student\1_modules1_5\baltimore_results). The baltimore_results folder contains an empty geodatabase named **baltresults.gdb** in which to save your data.

DATA (Current Workspace)   \student\1_modules1_5\baltimore_data
RESULTS (Scratch Workspace)   \student\1_modules1_5\baltimore_results

4. Save your map document to the \student\1_module1_5\baltimore_results folder. Name the file **baltimore_basics1**.

   Folder structure:

   EsriPress
      MSDLidar
         student
            1_modules1_5
               baltimore_data
                  LAS_files
                  baltimore.gdb
               baltimore_results
                  baltresults.gdb

Keeping track of where your data and results are located is always a challenge. In these activities, the path to access the data (workspace) and the path to store the results (scratch space) are listed at the bottom of each page. The directions will specify if results should be placed in the results folder or the results geodatabase.

5. Set environments:
   - Open the data frame properties. Set the map projection to Projected Coordinate Systems > State Plane > NAD_1983 (US Feet) > StatePlane_Maryland_FIPS_1900_Feet.
   - On the Geoprocessing menu, click Environments.
   - Expand Workspace, and set the current workspace to \student\1_modules1_5\ baltimore_data.
   - Set the scratch workspace to \student\1_modules1_5\baltimore_results.
   - For the output coordinate system, select Same as Display.

6. Connect to the baltimore_data folder.

In this folder, you will see a folder containing the LAS files for Baltimore that were downloaded from the US Geological Survey (USGS) Click website. When the LAS data was downloaded, the following metadata were provided:

- MD_Baltimore_2008
- NAD_1983_StatePlane_Maryland_FIPS_1900_Feet
- NAVD_1988_Foot
- Load Data 4/13/2009
- Tile 1 square mile
- 2008/MD_Baltimore_2008_1S1E
- 2008/MD_Baltimore_2008_1S1W
- 2008/MD_Baltimore_2008_2S1E

DATA (Current Workspace) \student\1_modules1_5\baltimore_data
RESULTS (Scratch Workspace) \student\1_modules1_5\baltimore_results

In the baltimore_data folder, you will also see baltimore.gdb, which contains a baltimore_layers feature dataset.

MAKING
SPATIAL
DECISIONS
USING GIS
AND LIDAR

1

BASIC LIDAR
TECHNIQUES

7. Right-click the baltimore_layers feature dataset and click Properties. Record the x,y coordinate system and z coordinate system.

Q1　**What are the x,y coordinate system and z coordinate system of the baltimore_layers feature dataset?**

Q2　**Do the coordinate systems for the feature dataset match the coordinate system of the LAS files?**

Analyze starting with a basemap

Part 1: Calculate statistics and build a LAS dataset

1. Add the water and baltimore_city layers from baltimore.gdb. Name the data frame **Baltimore**.

Lidar data and optional surface constraints can be added to a LAS dataset directly. A LAS dataset is not stored in a geodatabase; rather, a binary file is created and stored separately. Creating a LAS dataset is a quick process, as it is merely a pointer to the stored LAS files and surface constraints. The file extension that will be generated is .lasd.

Make sure the LAS files that you will use in the LAS dataset are all reasonably sized. A file size of 25 to 50 MB is recommended. LAS files should be no larger than 100 MB for optimal performance. The LAS files should contain no more than three million points per file when used in a LAS dataset.

2. Right-click the baltimore_data folder. Click New > LAS Dataset. Name the LAS dataset **baltimore_tiles**.

3. Double-click the baltimore_tiles.lasd to open the LAS dataset properties.

4. Click the LAS Files tab. Click Add Files and browse to the LAS_Files folder in baltimore_data. Add the four LAS files.

5. Click the Surface Constraints tab, and then add baltimore_city from the baltimore.gdb.

DATA (Current Workspace)　\student\1_modules1_5\baltimore_data
RESULTS (Scratch Workspace)　\student\1_modules1_5\baltimore_results

7

MAKING
SPATIAL
DECISIONS
USING GIS
AND LIDAR

BASIC LIDAR
TECHNIQUES

The surface feature type is Hard Clip. Surface constraint features are used to constrain feature-derived elevation values that represent surface characteristics in the LAS dataset. In this instance, the polygon baltimore_city defines the boundary of the LAS dataset.

6. Click the XY Coordinate System tab, and then choose NAD 1983 StatePlane Maryland FIPS 1900 (US Feet).

7. Click the Z Coordinate System tab, and then expand the Vertical Coordinate Systems > North America folders. Right-click NAVD 1988 and click Copy and Modify.
   - Type **NAVD_1988_foot** for the name.
   - In the Linear Unit section, click the Name arrow and select Foot.

8. Click OK. You have now constructed the LAS dataset. After you create the baltimore_tiles.lasd dataset, copy it and paste it into the baltimore_results folder.

Now that you have built the LAS dataset, you need to calculate statistics to examine the characteristics and quality of the data.

Point spacing is an indication of how close the data points are to each other. The point spacing determines the resolution of derived gridded products. The point spacing that is used in calculations is an average because point spacing is not consistent across the dataset. For example, tiles that have large bodies of water would have larger average point spacing because of the large homogenous features.

9. In the LAS Dataset Properties window, click the LAS Files tab. Use the information there to answer the following questions on your worksheet.

Q3 **What are the four point-spacing values given?**

Q4 **What is the average point spacing?**

Q5 **How does point spacing determine the resolution of derived gridded products?**

Q6 **How would the horizontal resolution of point spacing of a dataset affect the resolution of ground surface features?**

10. Click the Statistics tab, and then click Calculate to produce statistics for the LAS dataset.

MAKING
SPATIAL
DECISIONS
USING GIS
AND LIDAR

1

BASIC LIDAR
TECHNIQUES

The calculated statistics provide information about the lidar data.

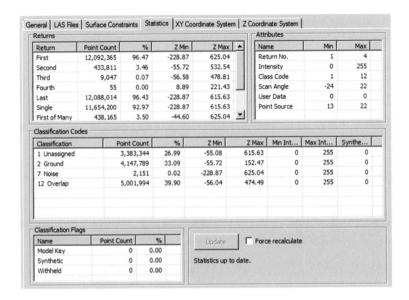

The statistics displayed include the available classification codes and return values as well as the range of elevation and intensity values. **Classification codes** indicate the type of object from which the pulse reflects. The most common type of classification is *bare earth ground*. The **return** indicates from what object the pulse is reflected (ground, buildings, low vegetation, and so on). Lidar systems can capture the first, second, third, and last return for each point. **Intensity** measures the strength of the pulse return, which is actually a measurement of how strongly the object reflects the laser pulse.

Q7   **What is the range of elevation?**

Q8   **If intensity is the strength of the return, and the value represents how strongly the object reflected light with the wavelength used by the laser system, what could cause the values to vary?**

Q9   **What are the classes given in this dataset?**

DATA (Current Workspace)   \student\1_modules1_5\baltimore_data
RESULTS (Scratch Workspace)   \student\1_modules1_5\baltimore_results

9

MAKING
SPATIAL
DECISIONS
USING GIS
AND LIDAR

1

BASIC LIDAR
TECHNIQUES

11. Before you close the LAS Dataset Properties window, click the General tab and select the "Store relative pathnames to data sources" check box.

Storing relative paths allows you to easily locate the LAS dataset and the LAS files in the file system.

12. Search for the LAS Dataset Statistics tool.

The LAS Dataset Statistics tool calculates statistics for a LAS dataset and generates a report as a text file, which can be given to the client. Saving the statistics as a text file is optional. It is adequate to just read the generated statistics.

13. Set the baltimore_tiles dataset as the input.

14. Output the file to the baltimore_results folder. Use the empty text file baltimore_statistics as the output file. Change the name of the empty text file to **baltimore_statistics2**.

This text file has all the pertinent information about the LAS dataset; however, it is not formatted in a useful way.

15. Start Excel and open baltimore_statistics2 (you may need to change the file type to All Files).

16. Select Delimited > Next.

17. Select Comma for Delimiters.

18. Select Next > Finish.

19. Save the file as **baltimore_statistics.xlsx**.

You have now created a LAS dataset statistics file for the client.

Deliverable 1: A LAS dataset statistics file.

HINT The LAS dataset statistics file provides information about the LAS files that is essential for understanding the lidar data with which you are working. Provide a short written analysis to explain the information generated in the statistics file.

DATA (Current Workspace)   \student\1_modules1_5\baltimore_data
RESULTS (Scratch Workspace)   \student\1_modules1_5\baltimore_results

The LAS dataset will display the extent of the files using a wireframe. All of the points cannot be displayed at the resolution of the wireframe because there are too many of them. Before zooming in to see the lidar points, investigate the wireframe area by adding some aerial imagery.

20. In ArcMap, zoom to baltimore_tiles.lasd. Click the arrow next to the Add Data button and click Add Basemap. Add the Imagery with Labels basemap, and then add the Topographic basemap.

Q10 **Describe the landscape in the four wireframe quadrants to someone who is unfamiliar with Baltimore.**

21. Turn off the basemaps, but do not remove them from the table of contents.

22. Zoom to the northwest quadrant of the map around the University of Maryland, Baltimore County, campus.

When you zoom to the LAS dataset around the University of Maryland, Baltimore County, you will see the lidar points. You will not see all of the lidar points (the data percentage shown in the table of contents gives the fraction of the points displayed). You can increase the number of points displayed.

23. Right-click baltimore_tiles.lasd and go to Properties. Click the Display tab and enter the following settings:
    - Increase the point limit to **5000000**.
    - Select the "Use scale to control full resolution" check box, and then change the full resolution scale value to **5000**.
    - Move the Point Density slider to Fine.
    - (Optional) Select the "Always display the LAS file extents" and "Display LAS file names" check boxes.

Q11 **What is the data percentage now shown in the table of contents?**

Q12 **What structures can you identify from the points in the four quadrants? Why are these structures easily identifiable?**

MAKING
SPATIAL
DECISIONS
USING GIS
AND LIDAR

1

BASIC LIDAR
TECHNIQUES

DATA (Current Workspace)   \student\1_modules1_5\baltimore_data
RESULTS (Scratch Workspace)   \student\1_modules1_5\baltimore_results

11

MAKING
SPATIAL
DECISIONS
USING GIS
AND LIDAR

BASIC LIDAR
TECHNIQUES

24. Create a presentation-quality map showing Baltimore City, water, and the wireframe of the extent of the LAS dataset. Pick an appropriate basemap to use as a background. Use correct cartographic principles.

25. Save the map document as **baltimore_basics1** to the baltimore_results folder.

26. Save the map document again as **baltimore_basics2**. (When you save the map document again as baltimore_basics2, it correctly saves documentation, the data frame projections, and the environment settings.)

Deliverable 2: A basemap showing the location of the LAS dataset.

Part 2: Use the interactive LAS Dataset toolbar

The interactive LAS Dataset toolbar offers a collection of tools that will work in the ArcMap or ArcScene applications. The toolbar allows you to display lidar points as a TIN-based surface or as a point cloud. It allows you to filter the points so you can see them as classes or returns. You can also produce slope and aspect surfaces, as well as elevation contours from the TIN. The toolbar provides a 2D profile viewer and a 3D viewer.

1. Turn on the LAS Dataset toolbar by going to the Customize menu, clicking Toolbars, and clicking LAS Dataset. You may dock the toolbar in any position. Notice that the baltimore_tiles.lasd is displayed in the LAS Dataset window.

When you added the LAS dataset to the table of contents, the points were displayed by elevation, and the filter was set to All, which meant all the points were shown.

You will now begin to investigate the different settings of the LAS Dataset toolbar. Before beginning your investigation, go to ArcGIS Help > Search. Use the ArcGIS Desktop Help Online if you have access to an Internet connection; otherwise, use the ArcGIS Desktop Help. Search for the "interactive LAS Dataset toolbar." The Help file gives a diagram of the toolbar and its tools. Take a few minutes to familiarize yourself with the toolbar.

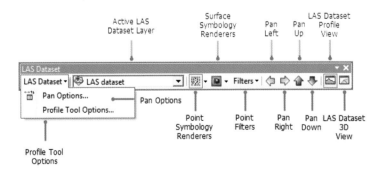

DATA (Current Workspace) \student\1_modules1_5\baltimore_data
RESULTS (Scratch Workspace) \student\1_modules1_5\baltimore_results

Explore point symbology renderers: elevation, class, and returns

Elevation displays the LAS points based on elevation.

MAKING
SPATIAL
DECISIONS
USING GIS
AND LIDAR

1

BASIC LIDAR
TECHNIQUES

2. Open baltimore_basics2.

3. Remove the baltimore_city and water layers.

4. Name the data frame **Point Symbology**.

5. Change the point symbology in the LAS Dataset toolbar to Elevation.

6. Zoom in and out around the LAS dataset.

7. Using the Measure tool, take several measurements between individual points. Set the unit to feet.

Q13 **What was the largest point spacing? What was the smallest point spacing?**

8. Zoom to various structures, such as docks, stadiums, buildings, and vegetation, and compare the lidar data to the Imagery basemap.

Q14 **Specifications for vertical and horizontal accuracy of lidar are designated in the requirements of the project. Vertical accuracy is usually greater than horizontal accuracy. Visually assess the horizontal accuracy of lidar points when compared to structures.**

Q15 **Why are the points not equally spaced?**

Class displays the LAS dataset based on the LAS classification code. This LAS dataset is classified using the following codes:
- 1. Unassigned
- 2. Ground
- 7. Noise
- 12. Overlap

DATA (Current Workspace)   \student\1_modules1_5\baltimore_data
RESULTS (Scratch Workspace)   \student\1_modules1_5\baltimore_results

13

MAKING
SPATIAL
DECISIONS
USING GIS
AND LIDAR

1

BASIC LIDAR
TECHNIQUES

9. Change the point symbology to Class.

Q16 ***What classification are the stripes of pink points running north to south through the dataset?*** Note: *In a future exercise, you will learn how to reclassify them.*

Return symbolizes the LAS dataset points by the lidar pulse return number.

10. Change the point symbology to Return.

Q17 ***How many returns are there, and what do they mean?***

11. Select an interesting area and change the point symbology back to Class. Again, notice that the buildings are gray, the ground is brown, and the overlay is pink.

DATA (Current Workspace)   \student\1_modules1_5\baltimore_data
RESULTS (Scratch Workspace)   \student\1_modules1_5\baltimore_results

Explore surface symbology renderers: elevation, aspect, slope, and contour

MAKING
SPATIAL
DECISIONS
USING GIS
AND LIDAR

1

*BASIC LIDAR
TECHNIQUES*

12. Copy the Point Symbology data frame and paste it into the table of contents. Change the name of the new data frame from Point Symbology to **Surface Symbology**.

Surface symbology displays the LAS dataset as a TIN-based surface. The four common surface models are elevation, aspect, slope, and contour.

13. Click the Surface Symbology arrow and choose Elevation.

Q18 **Describe how the data looks using this data structure.**

14. Display as Slope and then Aspect.

Q19 **Describe the difference between slope and aspect.**

15. Display as Contour.

Q20 **What do contours represent?**

16. Redisplay as Elevation.

DATA (Current Workspace)   \student\1_modules1_5\baltimore_data
RESULTS (Scratch Workspace)   \student\1_modules1_5\baltimore_results

15

MAKING
SPATIAL
DECISIONS
USING GIS
AND LIDAR

1

BASIC LIDAR
TECHNIQUES

Explore filters: all, ground, non ground, and first return

17. Copy the Surface Symbology data frame and paste it into the table of contents.

18. Name the data frame **Filters**.

There are four pre-set lidar filters in the menu:
- **All** uses all the lidar points.
- **Ground** uses only the lidar points designated as ground.
- **Non Ground** uses points that are not classified as ground.
- **First Return** uses only first return points.

19. When you copied the data frame, the filter was set to All. Change the filter to Ground.

Q21 *How does the dataset display change?*

20. Explore different ways to render the points or the TIN. Try different combinations of points, filters, and surfaces.

21. Reset the display to Elevation and the filter to Ground.

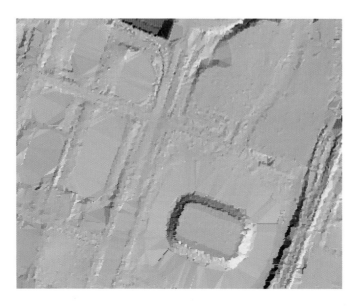

22. Create a presentation-quality map with three data frames: point class, elevation-TIN, and elevation with a ground filter. Use correct cartographic principles.

23. Save the map document as **baltimore_basics2** to the baltimore_results folder.

DATA (Current Workspace)   \student\1_modules1_5\baltimore_data
RESULTS (Scratch Workspace)   \student\1_modules1_5\baltimore_results

24. Save the map document again as **baltimore_basics3**. (When you save the map document again as baltimore_basics3, it correctly saves documentation, the data frame projections, and the environment settings.)

Deliverable 3: A map with three data frames showing points symbolized by class, an elevation TIN, and a TIN with a ground filter. A brief written analysis should be included on the map.

Explore the LAS Dataset toolbar profile view and 3D view

The interactive LAS Dataset toolbar provides a Profile View tool. The Profile View window allows the user to visualize, analyze, and edit lidar data using a 2D cross-sectional view. Visualizing cross sections of lidar data allows you to analyze collections of points from a unique perspective. It makes it easier to recognize particular features such as surface valleys, buildings, forest canopies, roads, and towers. Within the Profile View window, you can do the following:

- Measure distances and heights
- Manually edit and update LAS classification codes
- Select the LAS dataset 3D View window to visualize a selected portion of the LAS dataset

25. Open baltimore_basics3.

26. Switch to data view.

27. Remove the Point Symbology and Filters data frames.

28. Click the LAS Dataset Profile View button on the LAS Dataset toolbar.

29. Click a location to serve as a start point. Draw a straight horizontal or vertical line. Press the Shift key to straighten the line. A selection box displaying the profile appears. The dimensions and point count for the selection box are displayed in a small box in the ArcMap display window.

30. Construct several profiles.

MAKING
SPATIAL
DECISIONS
USING GIS
AND LIDAR

BASIC LIDAR
TECHNIQUES

DATA (Current Workspace)   \student\1_modules1_5\baltimore_data
RESULTS (Scratch Workspace)   \student\1_modules1_5\baltimore_results

17

Q22 **Explore different parts of the scene and use the Measure tool to obtain approximate heights of these objects:**

- Buildings
- Stadiums
- Houses
- Dock features
- Trees

31. Select an interesting profile and make a screen capture of it. Name the file **profile** and save it to the baltimore_results folder.

The LAS dataset 3D View window allows you to view your LAS dataset points or surface in 3D.

32. Zoom to the desired extent. Or, on the LAS Dataset toolbar, click the LAS Dataset 3D View button to view the scene in the 3D View window.

33. Make a screen capture of the 3D view and add it to a Word document. Name the file **viewer** and save it to the baltimore_results folder.

34. Insert the profile and viewer images into a Word document.
    - Write a brief description comparing the two visualizations.
    - Name the document **baltimore_visualization** and save it to the baltimore_results folder.
    - Alternatively, you can insert the two images as pictures into the baltimore_basics3 layout.

35. Save the map document as **baltimore_basics3**.

Deliverable 4: A document containing a profile, a 3D view, and a written description.

MAKING
SPATIAL
DECISIONS
USING GIS
AND LIDAR

1

*BASIC LIDAR
TECHNIQUES*

Draw conclusions and present the results

Now that you have explored the details of a lidar dataset and visualized it in two and three dimensions, choose a method of presenting your conclusions about Baltimore derived from lidar data. Always keep the audience in mind as you prepare to report your results; they may not share your GIS expertise. Your results can be presented in a Word document, a PowerPoint presentation, or a more technical presentation mode, such as ArcGIS Online.

DATA (Current Workspace)   \student\1_modules1_5\baltimore_data
RESULTS (Scratch Workspace)   \student\1_modules1_5\baltimore_results

19

MAKING
SPATIAL
DECISIONS
USING GIS
AND LIDAR

2

BASIC LIDAR
TECHNIQUES

PROJECT 2

San Francisco, California

Scenario

Repeat the analysis process used for the Baltimore, Maryland, lidar set.

Recommended deliverables

- **Deliverable 1:** A LAS dataset statistics file.
- **Deliverable 2:** A basemap showing the location of the LAS dataset.
- **Deliverable 3:** A map with three data frames showing points symbolized by class, an elevation TIN, and a TIN with a ground filter. A brief written analysis should be included on the map.
- **Deliverable 4:** A document containing a profile, a 3D view, and a written description.

The questions asked in this project are both quantitative and qualitative. They identify key points that should be addressed in your analysis and final presentation.

Document your work, set environments, and examine the data

1. Create your map document properties. Store relative pathnames to your data sources.
   Folder structure:
   EsriPress
       MSDLidar
           student
               1_modules1_5
                   sanfrancisco_data
                       LAS_files
                       sanfrancisco.gdb
                   sanfrancisco_results
                       sanfranresults.gdb

DATA (Current Workspace)   \student\1_modules1_5\sanfrancisco_data
RESULTS (Scratch Workspace)   \student\1_modules1_5\sanfrancisco_results

MAKING
SPATIAL
DECISIONS
USING GIS
AND LIDAR

BASIC LIDAR
TECHNIQUES

2. Set environments:
- Open the data frame properties. Set the map projection to Projected Coordinate Systems > UTM > NAD_1983 > Zone_10N.
- On the Geoprocessing menu, click Environments.
- Set the current workspace to \student\1_modules1_5\sanfrancisco_data.
- Set the scratch workspace to \student\1_modules1_5\sanfrancisco_results.
- For the output coordinate system, select Same as Display.

In the sanfrancisco_data folder, you will see a folder containing LAS files for San Francisco that were downloaded from the USGS Click website. The following metadata were provided with the files:
- ARRA-CA_SanFranCoast_2010
- NAD_1983_UTM_Zone_10N_Meters
- NAVD_1988_Meters
- Spring and Fall 2010
- Tile 1500 m by 1500 m
- ARRA-CA-SanFranCoast_2010_10SEG5282
- ARRA-CA-SanFranCoast_2010_10SEG5283
- ARRA-CA-SanFranCoast_2010_10SEG5382
- ARRA-CA-SanFranCoast_2010_10SEG5383

The sanfrancisco_data folder also contains the sanfrancisco.gdb file, which contains a layer dataset.

Q1   **What are the x,y coordinate system and z coordinate system of the san_franlayer dataset?**

Q2   **Do the coordinate systems for the feature dataset match the coordinate systems of the LAS files?**

Reminder: Before you close the LAS Dataset Properties window, click the General tab and select the "Store relative pathnames to the data source" check box. Storing relative paths allows you to easily relocate the LAS dataset and the LAS files in the file system.

DATA (Current Workspace)   \student\1_modules1_5\sanfrancisco_data
RESULTS (Scratch Workspace)   \student\1_modules1_5\sanfrancisco_results

21

2

Analyze starting with a basemap

Part 1: Calculate statistics and build a LAS dataset.

1. Add the counties and san_francisco layers from the sanfrancisco.gdb.

2. Build a LAS dataset using the LAS files in the LAS folder. Use san_francisco as the surface constraint.

3. Calculate statistics. Optionally, save them to the statistics.txt file in the sanfrancisco_results folder.

Q3 **What are the four point-spacing values given?**

Q4 **What is the average point spacing?**

Q5 **How does point spacing determine the resolution of derived gridded products?**

Q6 **How would the horizontal resolution of point spacing of a dataset affect the visibility of ground surface features?**

Q7 **What is the range of elevation?**

Q8 **If intensity is the strength of the return, and the value represents how strongly the object reflected the light used by the laser system, what could cause the values to vary?**

Q9 **What are the classes given in this dataset?**

4. Before you close the LAS Dataset Properties window, click the General tab and select the "Store relative pathnames to data sources" check box.

Storing relative paths allows you to easily relocate the LAS dataset and the LAS files in the file system.

5. Use the LAS Dataset Statistics tool and generate a report as a text file that can be imported into an Excel worksheet

Deliverable 1: A LAS dataset statistics file.

DATA (Current Workspace)   \student\1_modules1_5\sanfrancisco_data
RESULTS (Scratch Workspace)   \student\1_modules1_5\sanfrancisco_results

HINT The LAS dataset statistics file provides information about the LAS files that is essential for understanding the lidar data with which you are working. Provide a short written analysis to explain the information generated in the statistics file.

6. Add the Imagery with Labels and Topographic basemaps to sanfran.lasd.

Q10 ***Describe the landscape in the four wireframe quadrants to someone unfamiliar with San Francisco.***

7. Choose an interesting part of the lidar area and zoom to the LAS dataset. Increase the point limit to **5000000**, change the Full Resolution Scale value to **5000**, and move the Point Density slider to Fine.

Q11 ***What is the data percentage now shown in the table of contents?***

Q12 ***What structures can you now easily identify? Why?***

8. Create a presentation-quality map showing San Francisco and the wireframe of the extent of the LAS dataset. Pick an appropriate basemap to use as a background. Use correct cartographic principles.

9. Save the map document as **sanfran_basics1** to the sanfrancisco_results folder.

10. Save the map document again as **sanfran_basics2**.

Deliverable 2: A basemap showing the location of the LAS dataset.

Part 2: Use the interactive LAS Dataset toolbar.

Explore point symbology renderers: elevation, class, and returns

Elevation displays the LAS points based on elevation.

1. Remove any unnecessary layers from sanfran_basics2 and insert a new data frame. Name the data frame **Point Symbology**.

2. Using the Measure tool, take several measurements between points. Set the unit to feet.

Q13 ***What was the largest point spacing? What was the lowest point spacing?***

DATA (Current Workspace)   \student\1_modules1_5\sanfrancisco_data
RESULTS (Scratch Workspace)   \student\1_modules1_5\sanfrancisco_results

23

MAKING
SPATIAL
DECISIONS
USING GIS
AND LIDAR

2

BASIC LIDAR
TECHNIQUES

3. Zoom to various structures, such as docks, stadiums, buildings, and vegetation.

Q14 *Specifications for vertical and horizontal accuracy of lidar are designated in the requirements of the project. Vertical accuracy is usually greater than horizontal accuracy. Visually assess the horizontal accuracy of lidar points when compared to structures.*

Q15 *Why are the points not equally spaced?*

Class displays the LAS dataset based on the LAS classification code. This LAS dataset is classified using the following codes:

- 1. Unassigned
- 2. Ground
- 7. Noise
- 9. Water
- 10. Reserved

4. Change the point symbology to Class.

Q16 *What classification do blue points in the dataset represent?*

Return symbolizes the LAS dataset points by the lidar pulse return number.

5. Change the point symbology to Return.

Q17 *How many returns are there, and what do they mean?*

6. Select an interesting area and change the point symbology back to Class.

Explore surface symbology renderers: elevation, aspect, slope, and contour

7. Copy the Point Symbology data frame and paste it into the table of contents. Change the name of the new data frame from Point Symbology to **Surface Symbology**.

8. Display as Elevation.

Q18 *Describe how the data looks using this data structure.*

9. Display as Slope and then Aspect.

DATA (Current Workspace) \student\1_modules1_5\sanfrancisco_data
RESULTS (Scratch Workspace) \student\1_modules1_5\sanfrancisco_results

MAKING
SPATIAL
DECISIONS
USING GIS
AND LIDAR

2

BASIC LIDAR
TECHNIQUES

Q19 *Compare and contrast the slope and aspect surfaces. What does each show?*

10. Display as Contour.

Q20 *What do contours represent?*

11. Redisplay as Elevation.

Explore filters: all, ground, non ground, and first return

12. Copy the Surface Symbology data frame and paste it into the table of contents.

13. Name the data frame **Filters**. Change the filter to Ground.

Q21 *How does the dataset display change?*

14. Create a presentation-quality map with three data frames: point class, elevation-TIN, and elevation with a ground filter. Use correct cartographic principles.

15. Save the map document as **sanfran_basics2** to the sanfrancisco_results folder.

16. Save the map document again as **sanfran_basics3**.

Deliverable 3: A map with three data frames showing points symbolized by class, an elevation TIN, and a TIN with a ground filter. A brief written analysis should be included on the map.

Explore the LAS Dataset profile view and 3D view

17. Open sanfran_basics3 and remove the Point Symbology and Filters data frames.

18. Click the LAS Dataset Profile View button on the LAS Dataset toolbar. Construct several profiles and 3D views.

Q22 *Using the Measure tool, measure the approximate heights of the following objects:*
- Buildings
- AT&T Park
- Dock features
- Trees

DATA (Current Workspace)   \student\1_modules1_5\sanfrancisco_data
RESULTS (Scratch Workspace)   \student\1_modules1_5\sanfrancisco_results

MAKING
SPATIAL
DECISIONS
USING GIS
AND LIDAR

2

BASIC LIDAR
TECHNIQUES

19. Select an interesting profile and make a screen capture of it. Name the file **profile** and save it to the sanfrancisco_results folder.

The LAS dataset 3D View window allows you to view your LAS dataset points or surface in 3D.

20. Zoom to the desired extent. Or, click the LAS Dataset 3D View button to view the scene in the 3D View window.

21. Make a screen capture of the 3D view and add it to a Word document. Name the file **viewer** and save it to the sanfrancisco_results folder.

22. Insert the profile and viewer images into a Word document.
- Write a brief description of the two visualizations.
- Name the document **sanfrancisco_visualization** and save it to the sanfrancisco_results folder.
- Alternatively, you can insert the two images as pictures into the sanfran_basics3 layout.

Deliverable 4: A document containing a profile, a 3D view, and a written description.

Draw conclusions and present the results

Now that you have explored the details of a lidar dataset and visualized it in two and three dimensions, choose a method of presenting your conclusions about San Francisco derived from lidar data. Always keep the audience in mind as you prepare to report your results; they may not share your GIS expertise. Your results can be presented in a Word document, a PowerPoint presentation, or a more technical presentation mode, such as ArcGIS Online.

DATA (Current Workspace)   \student\1_modules1_5\sanfrancisco_data
RESULTS (Scratch Workspace)   \student\1_modules1_5\sanfrancisco_results

MAKING
SPATIAL
DECISIONS
USING GIS
AND LIDAR

3

*BASIC LIDAR
TECHNIQUES*

PROJECT (3)

On your own

Scenario

You have worked through a guided project and repeated the analysis for another project. In this project, you will reinforce your skills by researching and analyzing a similar scenario entirely on your own. First, identify your study area and acquire data for your analysis. You may want to study a local area.

Here is a list of topics that have been studied using lidar. The following represent possible project ideas:

- Forest characterization—canopy height and density
- Flood modeling
- Finding faults
- Geomorphic mapping
- Stream slope
- Archaeology field campaigns
- Mining—calculation of ore volumes
- Wind farm optimization

MAKING
SPATIAL
DECISIONS
USING GIS
AND LIDAR

3

BASIC LIDAR
TECHNIQUES

Many different websites distribute lidar data. Most of the available lidar data is for locations within the United States. There are national data sites, and there are state data distributors. Reading the metadata and identifying a specific area of study is critical before downloading files. Some lidar files come compressed and require third-party software to convert them into the LAS format used by ArcMap.

- Open Topography facilitates community access to high-resolution, Earth science-oriented topography data. It is a National Science Foundation-funded data facility. http://www.opentopography.org.
- The Earth Explorer provides online access to remotely sensed data from the US Geological Survey Earth Resources Observation and Science (EROS) Center archive. http://earthexplorer.usgs.gov/.
- National Oceanic and Atmospheric Administration, a world leader in coast science and management, provides state lidar datasets. http://www.csc.noaa.gov/dataviewer/#.
- Wikipedia lists national lidar datasets, organized according to state. http://en.wikipedia.org/wiki/National_LIDAR_Dataset_%E2%80%93_USA.

Research

Research the problem and answer the following questions:

1. What is the area of study?
2. What problem are you going to study?
3. What data is available?

2D AND 3D MODELS (RASTER AND TERRAIN)

SCENARIO ···

It is now feasible to model urban landscapes at both the city and building level. 2D and 3D urban models can show detailed terrain, buildings, streets, and urban vegetation. Models using both raster and terrain formats are available to city planners. These models can be used to evaluate space and to simulate building plans to inform local communities. Baltimore and San Francisco have asked your GIS Company to provide them with models of their cities in both raster and terrain data formats.

Raster or gridded elevations models can be made from LAS datasets. The LAS data can be broken into different segments based on the returns. The most common segments are *ground* and *first return*. Ground represents bare earth or surface topography, and first return typically includes buildings and tree canopies. Ground is often referred to as DEM (digital elevation model) and first return as DSM (digital surface model).

Terrains are produced by using a combination of points and breaklines to produce a series of triangulated irregular networks (TINs), each of which has its own scale range. The use of terrains as a data storage and visualization method enables faster viewing than other elevation data types. Storing surface information as feature classes in a geodatabase is one of the benefits of creating a terrain dataset. Breaklines usually represent lakes, shorelines, and large rivers, or they are used to delineate a study area. Both data formats have pros and cons. City officials would like you to assess the different models and advise them on appropriate uses.

Projects included in this module

- **Project 1:** Baltimore, Maryland
- **Project 2:** San Francisco, California
- **Project 3:** On your own

Student worksheets

Project 1: Baltimore, Maryland

- File name: 2a_DEM_DSM_Terrains_worksheet.docx
- Location:EsriPress\MSDLidar\student\1_modules1_5\2DEM_DSM_Terrains\project1_documents

Project 2: San Francisco, California

- File name: 2b_DEM_DSM_Terrains_worksheet.docx
- Location:EsriPress\MSDLidar\student\1_modules1_5\2DEM_DSM_Terrains\project2_documents

Objectives

- Create a geodatabase and a feature dataset.
- Convert a LAS dataset into a raster.
  - DEM
  - DSM
  - Convert 2D buildings into 3D buildings.
  - Convert streets to elevated shapefiles.
- Convert a point cloud into a terrain.
  - Incorporate breaklines.
- Compare the models.

PROJECT (1)

Baltimore, Maryland

Lidar data for the Baltimore footprints was obtained from the OpenBaltimore website: https://data.baltimorecity.gov/Geographic/Building-Footprint-Shape/deus-s85f.

Recommended deliverables

- **Deliverable 1:** A document with a 3D view of the LAS dataset.
- **Deliverable 2a:** A map containing data frames of the DEM and of the DSM.
- **Deliverable 2b:** A 3D view of the DEM with streets and buildings.
- **Deliverable 3:** A map containing the All Terrain, DEM Terrain, and DSM Terrain data frames.
- **Deliverable 4:** A comparison of the three models. Both datasets have pros and cons. Your clients would like you to assess the different 3D models and offer advice on how best to use them.

The questions asked in this project are both quantitative and qualitative. They identify key points that should be addressed in your analysis and final presentation.

Document your work, set environments, and examine the data

In this exercise, you will be introduced to the ArcScene application. ArcScene is a 3D viewer that allows you to display, analyze, and interact with your 3D or 2D data in a 3D space. ArcScene allows you to view LAS datasets, rasters, and 3D features. ArcScene does not allow you to view terrain datasets.

1. Open ArcScene.

2. Add relevant scene document properties. Store relative pathnames to your data sources.

DATA (Current Workspace)   \student\1_modules1_5\baltimore_data
RESULTS (Scratch Workspace)   \student\1_modules1_5\baltimore_results

MAKING
SPATIAL
DECISIONS
USING GIS
AND LIDAR

1

2D AND
3D MODELS
(RASTER AND
TERRAIN)

The folder structure and environments for module 2 are the same as those in module 1.

3. Save the map document as **balt _3D1** to the baltimore_results folder.

Analyze

View a LAS dataset in 3D

1. Add the baltimore_tiles.lasd dataset.

2. Right-click Scene layers. Click Scene Properties. On the General tab, set the vertical exaggeration to Calculate From Extent.

Q1 ***Describe what you visualize when looking at the LAS dataset in 3D and how the dataset is displayed.***

You can turn on the LAS Dataset toolbar in ArcScene. You immediately see that there are fewer options available than in ArcMap. However, you can still change the point symbology, the surface symbology, and the filters.

3. Change the filter to Ground, to Non Ground, and then back to All.

Q2 ***How does changing the filter change the appearance of the dataset?***

4. To interact with the LAS dataset, do the following:
   - Right-click the baltimore_tiles.lasd, and then click Properties. Click the Symbology tab.
   - Click Add.
   - Select the "Contour with the same symbol" option.
   - Leave the contour interval as 5.
   - Click OK.

Q3 ***What does the addition of the contour lines do to the display of the data?***

5. Make a screen capture of the LAS dataset in 3D showing elevation and one showing contours. Insert the images in a Word document titled "2D_3D models" and save it to the 1_modules1_5\2DEM_DSM_Terrains\project1_documents folder.

MAKING
SPATIAL
DECISIONS
USING GIS
AND LIDAR

1

*2D AND
3D MODELS
(RASTER AND
TERRAIN)*

6. Save the project as **balt_3D1.sxd**. Save the project again as **balt_3D2.sxd**.

Deliverable 1: A document with a 3D view of the LAS dataset.

DATA (Current Workspace)   \student\1_modules1_5\baltimore_data
RESULTS (Scratch Workspace)   \student\1_modules1_5\baltimore_results

35

MAKING
SPATIAL
DECISIONS
USING GIS
AND LIDAR

1

2D AND
3D MODELS
(RASTER AND
TERRAIN)

Construct a DEM dataset

1. Open ArcMap.

2. Add map document properties. Store relative pathnames to your data sources.

The folder structure and environments for this module are the same as those in the previous module.

3. Save the map document as **baltimore_models1** to the baltimore_results folder.

4. Name the data frame **Baltimore DEM**.

5. Add the baltimore_city and water layers.

6. Add the baltimore_tiles.lasd dataset.

Zoom around the LAS dataset.

04   *Where are there no lidar points? Why?*

In order to change the LAS dataset to a raster, you must exclude the water areas that have no lidar points.

7. Remove the baltimore_tiles.lasd dataset.

8. In ArcCatalog, right-click the dataset and go to Properties.

9. Click the Surface Constraints tab and add the baltimore_city layer. Set the surface feature type to Hard Clip.

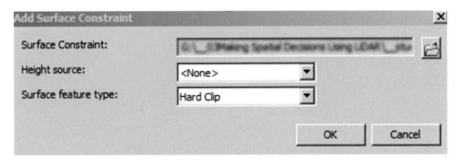

DATA (Current Workspace)   \student\1_modules1_5\baltimore_data
RESULTS (Scratch Workspace)   \student\1_modules1_5\baltimore_results

10. Add the baltimore_tiles.lasd dataset back to the table of contents.

In order to make a bare earth model or DEM, you need to filter the LAS dataset so that only the points that represent the ground are selected.

11. On the LAS Dataset toolbar, set the filter to Ground.

Q5   ***How does that change affect the look of the LAS dataset?***

You are now ready to convert the LAS dataset to a raster. To make this conversion, ArcMap needs to interpolate between the LAS dataset points and create a raster where each pixel has a value representing the average elevation in that area.

12. Open the LAS Dataset to Raster tool.

The LAS Dataset to Raster tool allows you to convert LAS datasets to a raster surface model for additional analysis. You can create a raster using elevation, intensity, or RGB (red, green, blue) values stored in the lidar files (*.las) referenced by the LAS dataset. In this exercise, elevation is the value needed to perform the analysis.

13. Enter the following parameters:
    - Set the input LAS dataset to baltimore_tiles.lasd.
    - Set the output raster to student\1_modules1_5\baltimore_results\baltresults.gdb.
    - Set the value field to ELEVATION.
    - Binning represents the interpolation method used to produce the raster. Enter the following binning settings:
        - Set the cell assignment type to MAXIMUM.
        - Set the void fill method to NATURAL_NEIGHBOR.
    - Set the output data type to INT (for integer).
    - Set the sampling type to CELLSIZE.
    - Set the sampling value to **10**.
    - Click OK.

14. Classify the raster and select an elevation color ramp. Format the cells so that there are no decimals.

Q6   ***What are the lowest and highest elevations in the DEM?***

DATA (Current Workspace)   \student\1_modules1_5\baltimore_data
RESULTS (Scratch Workspace)   \student\1_modules1_5\baltimore_results

37

MAKING
SPATIAL
DECISIONS
USING GIS
AND LIDAR

1

2D AND
3D MODELS
(RASTER AND
TERRAIN)

Q7   *Are there any areas with an elevation below sea level? Where?*

Q8   *Describe one immediate benefit of using the baltimore_DEM dataset.*

Construct a DSM dataset

1. Copy the Baltimore DEM data frame and paste it into the table of contents. Name the new data frame **Baltimore DSM**.

2. Remove baltimore_DEM.

3. To construct a DSM, repeat steps 10 and 11 of the previous section. However, instead of filtering for ground, filter for non ground. Name the file **baltimore_DSM** and save it to the baltresults.gdb.

4. Classify the raster and select an appropriate color ramp.

Q9   *What are the lowest and highest elevations in the DSM?*

5. Construct a map with two data frames that show the DEM and the DSM of the Baltimore tiles. Use appropriate cartographic principles.

6. Save the map document as **baltimore_models1**. Save the map document again as **baltimore_models2**.

Q10   *Compare and contrast the DEM and DSM. For what purpose might each format be useful?*

Deliverable 2a: A map containing data frames of the DEM and the DSM.

Q11   *Write a paragraph describing the processes used to create a DEM and a DSM. This analysis should focus on the underlying processes of the software.*

DATA (Current Workspace)   \student\1_modules1_5\baltimore_data
RESULTS (Scratch Workspace)   \student\1_modules1_5\baltimore_results

Construct a 3D raster model

1. Open balt_3D2.sxd.

2. Remove baltimore_tiles.lasd.

3. Right-click Scene Layers, and go to Scene Properties. On the General tab, set the vertical exaggeration to None.

4. Add baltimore_DEM from the baltresults.gdb.

5. Right-click baltimore_DEM, and go to Properties. On the Base Heights tab, choose the "Floating on a custom surface" option. Choose an appropriate color ramp.

6. Add baltimore_DSM.

7. Right-click baltimore_DSM, and go to Properties. On the Base Heights tab, choose the "Floating on a custom surface" option. Choose baltimore_DSM. Choose an appropriate color ramp.

8. Right-click Scene Layers, and go to Scene Properties. On the General tab, set the vertical exaggeration to Calculate from Extent.

Q12   *Describe the 3D raster DEM and DSM visualization.*

A better 3D visualization could be made by using the raster baltimore_DEM as the base and adding vector highways and building footprints.

9. Remove baltimore_DSM.

10. Right-click Scene Layers, and go to Scene Properties. On the General tab, set the vertical exaggeration to Calculate from Extent.

11. Add highways from baltimore_data\baltimore.gdb\baltimore_layers.

12. Use the Interpolate Shape tool to derive z-values (heights) from the elevation raster.

The Interpolate Shape tool interpolates z-values for a feature class based on elevation derived from a raster.

DATA (Current Workspace)   \student\1_modules1_5\baltimore_data
RESULTS (Scratch Workspace)   \student\1_modules1_5\baltimore_results

39

MAKING
SPATIAL
DECISIONS
USING GIS
AND LIDAR

1

2D AND
3D MODELS
(RASTER AND
TERRAIN)

13. Enter the following parameters:
- Set the input surface to baltimore_DEM.
- Set the input feature class to highways.
- Set the output feature class to baltresults.gdb\baltimore_layers**highways3D.**

14. Right-click highways3D, and go to Layer Properties. On the Base Heights tab, choose the "Floating on a custom surface" option. Choose baltimore_DEM.

15. Make highways3D an appropriate color with a 3-point width.

16. Remove highways.

17. Save the ArcScene map document as **balt_3D2.sxd**.

The next part of the exercise involves extracting the height of buildings from the baltimore_DSM surface. It is a process that is more easily done in ArcMap. You will first change the building polygons to points. Then, you will obtain heights from the DSM you created earlier. Finally, you will add those heights to the building layer.

18. Open ArcMap.

19. From baltimore_data\baltimore.gdb\baltimore_layers, add the bldgs layer.

If you open the attribute table for bldgs, you will see that there is no height value included.

20. Right-click bldgs, and go to Data > Export Data. Output the feature class to baltresults.gdb\baltimore_layers. Name the file **bldgs2**.

21. Remove bldgs.

The Feature to Point tool creates a feature class containing points generated from the representative locations of input features. The input surface for this exercise is the raster baltimore_DSM, which represents the digital surface model raster.

22. Run the Feature to Point tool. Enter the following parameters:
- Set the input to bldgs2.
- Set the output feature class to \baltresults.gdb\baltimore_layers**bldgs_feature_to_point**.

23. Add baltimore_DSM.

You will use this surface to obtain the height values of the buildings.

24. Search for and open the Add Surface Information tool. Enter the following parameters:
- Set the input feature class to bldgs_feature_to_point.
- Set the input surface to baltimore_DSM.
- Select Z.
- Click OK.

To add the building heights to bldgs2, you must join bldgs_feature_to_point to bldgs2.

25. Right-click bldgs2, go to Join and Relates, and choose Join.

26. Choose OBJECTID_12, and it will automatically choose the OBJECTID_12 field for bldgs_feature_to_point.

27. Click OK.

28. Right-click bldgs2, and go to Data > Export Data. Save the file to baltresults.gdb\layers**bldgs3**.

29. Close ArcMap. You do not have to save the map document.

30. Open balt_3D2.sxd.

31. Add bldgs3. Right-click bldgs3, and go to Properties. On the Base Heights tab, make the elevation from surface floating on the DEM.

32. Right-click bldgs3, and go to Properties. Click the Extrusion tab. Extrude by z-value.

33. Right-click Scene layers and calculate the vertical exaggeration.

34. Save the project as **balt_3Ds.sxd**.

35. Make a screen capture of your raster and building 3D model. Insert the image into the "2D_3D models" document in the 1_modules1_5\2DEM_DSM_Terrains\project1_documents folder.

DATA (Current Workspace)   \student\1_modules1_5\baltimore_data
RESULTS (Scratch Workspace)   \student\1_modules1_5\baltimore_results

41

MAKING
SPATIAL
DECISIONS
USING GIS
AND LIDAR

1

*2D AND
3D MODELS
(RASTER AND
TERRAIN)*

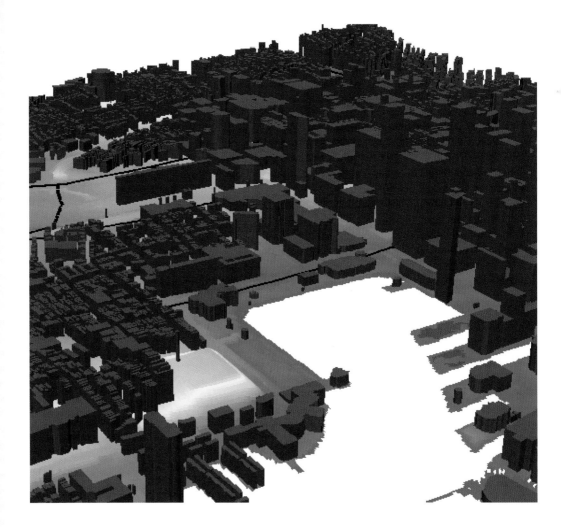

Q13 **Describe the 3D raster/vector model.**

Deliverable 2b: A 3D view of the DEM with streets and buildings.

Construct a terrain model

A terrain dataset is a TIN-based surface built from measurements stored as features in a geodatabase. A terrain dataset is a good way to manage a large collection of points, such as lidar data. However, you cannot view terrains in 3D.

1. Open baltimore_models2.mxd.

2. Remove Baltimore DSM. Also remove water and the DEM from the Baltimore DEM frame. Name the data frame **All Terrain**.

DATA (Current Workspace)   \student\1_modules1_5\baltimore_data
RESULTS (Scratch Workspace)   \student\1_modules1_5\baltimore_results

Remember: A terrain dataset is a multiresolution, TIN-based surface built from measurements stored in a geodatabase. They are typically made from lidar and sonar. Terrains reside in the geodatabase, inside feature datasets with the features used to construct them.

To create a terrain dataset, you must convert the LAS files to points. As with the raster displays, the points can be filtered to All, Ground, and Non Ground.

The LAS to Multipoint tool imports one or more files in LAS format into a new multipoint feature class. The LAS points can be classified into a number of categories, including ground or non ground.

MAKING
SPATIAL
DECISIONS
USING GIS
AND LIDAR

2D AND
3D MODELS
(RASTER AND
TERRAIN)

3. Run the LAS to Multipoint tool. Enter the following parameters:
   • Input the four LAS files from the baltimore_data folder.
   • Set the ouput feature class to baltresults.gdb\terrain_layers**ptcloudall.**
   • Set the average point spacing to **3**.
   • Set the coordinate system to NAD_1983_StatePlane_Maryland_FIPS_1900_Feet.

4. Repeat step 3. Choose **1** for the class code, which represents all the non ground points. Name the file **ptclouddsm**.

5. Repeat step 3 again. Choose **2** for the class code, which represents all the ground points. Name the file **ptclouddem**.

6. Open the attribute table of each raster. Remember: Point Count is the number of points per value. Right-click PointCount and look at the statistics.

Q14 ***How many points are in ptcloudall, ptclouddsm, and ptclouddem?***

Now you must turn each of these point cloud datasets into terrains.

7. In baltresults.gdb, right-click terrain_layers. Go to New Terrain.

8. Add baltimore_city to the table of contents. Right-click baltimore_city and go to Data > Export Data. Export the data to the terrain_all_terrain_baltimore feature dataset. Enter the following settings:
   • Name the terrain **all_terrain_baltimore**.
   • Choose ptcloudall and baltimore_city.
   • Set the average spacing to **3**.
   • Set baltimore_city as a hard_clip.

DATA (Current Workspace)   \student\1_modules1_5\baltimore_data
RESULTS (Scratch Workspace)   \student\1_modules1_5\baltimore_results

43

- Accept defaults for select pyramid type.
- Calculate pyramid properties.
- Finish and build the terrain.
- Zoom to an interesting area.

9. Insert a new data frame and name it **terrain DEM**.

10. Repeat step 8 using ptclouddem. Name the terrain **terrainDEM**.

11. Insert a new data frame and name it **terrain DSM**.

12. Repeat step 8 using ptclouddsm. Name the terrain **terrainDSM**.

13. Construct a map with three data frames that show the differences in the terrain models.

Cartographic tip: The use of bookmarks

You can use bookmarks to ensure all three of the data frames have the same perspective.
- Click the data frame that has the desired perspective.
- Go to the Bookmarks menu. Create a bookmark and name it.
- Go to Manage Bookmarks and save the bookmark to your baltimore_results folder. Name the file **3_data_frame_baltimore**.
- Click the two other data frames and zoom to the 3_data_frame_baltimore bookmark.

14. Save the map document as **baltimore_models2.mxd**.

Deliverable 3: A map showing the All Terrain, DEM Terrain, and DSM Terrain data frames.

Compare the 2D and 3D models

Deliverable 4: A comparison of the three models. Both datasets have pros and cons. Your clients would like you to assess the different 3D models and offer advice on how best to use them.

Draw conclusions and present the results

In this activity, you have created many different models for Baltimore to compare surfaces created from lidar data. Choose a method of presenting your conclusions about the strengths and weaknesses of the various methods as well as the most effective use of each. Always keep the audience in mind as you prepare to report your results; they may not share your GIS expertise. Your results can be presented in a Word document, a PowerPoint presentation, or a more technical presentation mode, such as ArcGIS Online.

DATA (Current Workspace)   \student\1_modules1_5\baltimore_data
RESULTS (Scratch Workspace)   \student\1_modules1_5\baltimore_results

45

MAKING
SPATIAL
DECISIONS
USING GIS
AND LIDAR

2

2D AND
3D MODELS
(RASTER AND
TERRAIN)

PROJECT 2

San Francisco, California

Lidar data was obtained from the US Geological Survey Earth Explorer website: http://earthexplorer.usgs.gov/.

Building footprints were obtained from the SF OpenData website: https://data.sfgov.org/Other/Data-Catalog/h4ui-ubbu.

Recommended deliverables

- **Deliverable 1:** A document with a 3D view of the LAS dataset.
- **Deliverable 2a:** A map containing data frames of the DEM and of the DSM.
- **Deliverable 2b:** A 3D view of the DEM with streets and buildings.
- **Deliverable 3:** A map containing the All Terrain, DEM Terrain, and DSM Terrain data frames.
- **Deliverable 4:** A comparison of the three models. Both datasets have pros and cons. Your clients would like you to assess the different 3D models and offer advice on how best to use them.

The questions asked in this project are both quantitative and qualitative. They identify key points that should be addressed in your analysis and final presentation.

DATA (Current Workspace)   \student\1_modules1_5\sanfrancisco_data
RESULTS (Scratch Workspace)   \student\1_modules1_5\sanfrancisco_results

Document your work, set environments, and examine the data

MAKING
SPATIAL
DECISIONS
USING GIS
AND LIDAR

2

2D AND
3D MODELS
(RASTER AND
TERRAIN)

1. Open ArcScene.

2. Add scene document properties. Store relative pathnames to your data sources.

The folder structure and environments are the same as those in module 1.

3. Save the map document as **SF_3D1** to the sanfrancisco_results folder.

Analyze

View a LAS dataset in 3D

1. Add the sanfran_tiles.lasd dataset and calculate the vertical exaggeration.

Q1   ***Describe what you visualize when looking at the LAS dataset in 3D and how the dataset is displayed.***

You can turn on the LAS Dataset toolbar in ArcScene. You immediately see that there are fewer options available than in ArcMap. However, you can still change the point symbology, the surface symbology, and the filters.

2. Change filters.

Q2   ***How does changing the filter change the look of the dataset?***

3. Create contours.

Q3   ***What does the addition of the contour lines do to the display of the data?***

4. Make a screen capture of the LAS dataset in 3D showing elevation and one showing contours. Insert the images in a Word document titled "2D_3D models" and save it to the 1_modules1_5\2DEM_DSM_Terrains\project2_documents folder.

DATA (Current Workspace)   \student\1_modules1_5\sanfrancisco_data
RESULTS (Scratch Workspace)   \student\1_modules1_5\sanfrancisco_results

47

MAKING
SPATIAL
DECISIONS
USING GIS
AND LIDAR

2

2D AND
3D MODELS
(RASTER AND
TERRAIN)

Deliverable 1: A document with a 3D view of the LAS dataset.

DATA (Current Workspace)   \student\1_modules1_5\sanfrancisco_data
RESULTS (Scratch Workspace)   \student\1_modules1_5\sanfrancisco_results

Construct a DEM dataset

1. Open ArcMap and add appropriate map document properties. Store relative pathnames to your data sources.

The folder structure and environments for module 2 are the same as those in module 1.

2. Save the map document as **SF_models1** to the sanfrancisco_results folder. Save the map document again as **SF_models2**.

3. Name the data frame **SF_DEM** and add the sa_sanfrancisco layer and sanfran.lasd.

4. Construct a DEM raster dataset. Name it **SF_DEM** and save it to sanfrancisco_results\ sanfrancisco.gdb.

MAKING
SPATIAL
DECISIONS
USING GIS
AND LIDAR

2

2D AND
3D MODELS
(RASTER AND
TERRAIN)

Q4   ***Where is there no lidar coverage? Why?***

5. Remember to use the following parameters:
   - Set the filter in the LAS dataset to show only ground points.
   - Use sa_sanfrancisco as a hard clip.
   - The point spacing is 1, so the cell size should be **4**.
   - Other parameters should be the same as for the Baltimore conversion.
   - Save the file as **SF_DEM** to sanfrancisco_results\sanfrancisco.gdb.

Q5   ***How does filtering the LAS dataset affect its appearance?***

Q6   ***What are the lowest and highest elevations in the DEM?***

Q7   ***Is there any elevation that is below sea level? Where?***

Q8   ***Describe one immediate benefit of using the SF_DEM dataset.***

DATA (Current Workspace)   \student\1_modules1_5\sanfrancisco_data
RESULTS (Scratch Workspace)   \student\1_modules1_5\sanfrancisco_results

49

MAKING
SPATIAL
DECISIONS
USING GIS
AND LIDAR

2

2D AND
3D MODELS
(RASTER AND
TERRAIN)

Construct a DSM dataset

1. Make a new data frame named **SF_DSM** and repeat the process previously described. Remember to set the filter to non ground. Save the file as **SF_DSM**.

Q9   *What are the lowest and highest elevations in the DSM?*

2. Construct a map with two data frames that show the DEM and the DSM of the San Francisco tiles. Use appropriate cartographic principles.

3. Save the map document as **SF_model1**. Save the map document again as **SF_model2**.

Deliverable 2a: A map containing data frames of the DEM and the DSM.

Construct a 3D raster model

1. Open sf_3D2.sxd and reset the vertical exaggeration to none.

2. Add the SF_DEM and the SF_DSM. Adjust the base height and vertical exaggeration.

Q10   *Describe the 3D raster DEM and DSM visualization.*

A better 3D visualization could be made by using the raster baltimore_DEM as the base and adding vector highways and building footprints. The San Francisco buildings dataset already includes the maximum height of the buildings; therefore, you only have to derive the z-values from the SF_DEM for the highways.

3. Use the Interpolate Shape tool and save the file as **highways3D** to sanfrancisco_results.gdb.

4. Add highways3D and obtain the base height.

5. Add bldgs, extrude by maxheight, and reset the vertical exaggeration.

6. Save the ArcScene map document as **SF_3d2.sxd**.

7. Make a screen capture of your raster and building 3D model. Insert the image into the "2D_3D models" document in the 1_modules1_5\2DEM_DSM_Terrains\project2_documents folder.

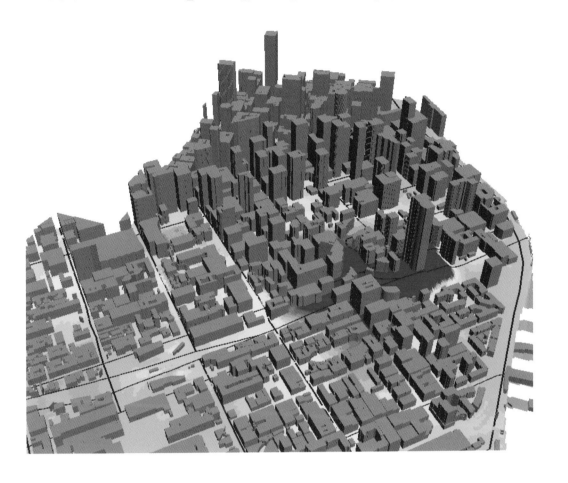

Q11 **Describe the 3D raster/vector model.**

Deliverable 2b: A 3D view of the DEM with streets and buildings.

DATA (Current Workspace)   \student\1_modules1_5\sanfrancisco_data
RESULTS (Scratch Workspace)   \student\1_modules1_5\sanfrancisco_results

51

MAKING
SPATIAL
DECISIONS
USING GIS
AND LIDAR

2

2D AND
3D MODELS
(RASTER AND
TERRAIN)

Construct a terrain model

1. Open SF_model2D2.mxd.

2. Remove all layers.

3. Create sfptall, sfptdem, and sfptdsm using the LAS to Multipoint tool. Save to the sanfrancisco.gdb\terrain_layers folder.

Remember: 1 is non ground, and 2 is ground. Average point spacing is 1.

Q12 ***How many points are in sfptall, sfptdem, and sfptdsm?***

4. Make a terrain from each of the point clouds. Name the terrains **all_terrain**, **DSM_terrain**, and **DEM_terrain**, respectively. Save them to sanfrancisco.gdb\terrain_layers.

Remember: Use sa_sanfrancisco as a hard_clip, with 1 as the average spacing.

5. Construct a map with three data frames that show the differences in the terrain models.

6. Save the map document as **SF_model2D2.mxd**.

Deliverable 3: A map showing the All Terrain, DEM Terrain, and DSM Terrain data frames.

Compare the 2D and 3D models

Deliverable 4: A comparison of the three models. Both datasets have pros and cons. Your clients would like you to assess the different 3D models and offer advice on how best to use them.

Draw conclusions and present the results

In this activity, you have created many different models for San Francisco to compare surfaces created from lidar data. Choose a method of presenting your conclusions. Always keep the audience in mind as you prepare to report your results; they may not share your GIS expertise. Your results can be presented in a Word document, a PowerPoint presentation, or a more technical presentation mode, such as ArcGIS Online.

DATA (Current Workspace)   \student\1_modules1_5\sanfrancisco_data
RESULTS (Scratch Workspace)   \student\1_modules1_5\sanfrancisco_results

MAKING
SPATIAL
DECISIONS
USING GIS
AND LIDAR

3

*2D AND
3D MODELS
(RASTER AND
TERRAIN)*

PROJECT (3)

On your own

Scenario

You have worked through a guided project and repeated the analysis for another project. In this project, you will reinforce your skills by researching and analyzing a similar scenario entirely on your own. First, identify your study area and acquire data for your analysis. You may want to study a local area.

Here is a list of topics that have been studied using lidar. The following represent possible project ideas:

- Forest characterization—canopy height and density
- Flood modeling
- Finding faults
- Geomorphic mapping
- Stream slope
- Archaeology field campaigns
- Mining—calculation of ore volumes
- Wind farm optimization

MAKING
SPATIAL
DECISIONS
USING GIS
AND LIDAR

3

2D AND
3D MODELS
(RASTER AND
TERRAIN)

Many different websites distribute lidar data. Most of the available lidar data is within the United States. There are national data sites, and there are state data distributors. Reading the metadata and identifying a specific area of study is critical before downloading files. Some lidar files come compressed and require third-party software to convert them into the LAS format used by Esri.

- Open Topography facilitates community access to high-resolution, Earth science-oriented, topography data. It is a National Science Foundation-funded data facility. http://www.opentopography.org.
- The Earth Explorer provides online access to remotely sensed data from the US Geological Survey Earth Resources Observation and Science (EROS) Center archive. http://earthexplorer.usgs.gov/.
- National Oceanic and Atmospheric Administration, a world leader in coast science and management, provides state lidar datasets. http://www.csc.noaa.gov/dataviewer/#.
- Wikipedia lists national lidar datasets, organized according to state. http://en.wikipedia.org/wiki/National_LIDAR_Dataset_%E2%80%93_USA.

Research

Research the problem and answer the following questions:

1. What is the area of study?
2. What problem are you going to study?
3. What data is available?

VOLUMETRIC ANALYSIS AND SHADOW MAPS

SCENARIO ···

Baltimore and San Francisco have limited surface parking and need to explore underground options. City officials recently attended a GIS conference where new 3D tools for volumetric analysis using high-accuracy lidar were demonstrated. The cities would like your company to evaluate new buildings under development and give an estimate of how much land will have to be excavated for underground parking.

The analysis also involves the use of volumetric shadow analysis of the new structures to show areas of shadow in an adjoining park. The city planners want to incorporate eating areas with benches at designated sites and have asked for a shadow map showing areas of shadow cast by the buildings at 10 a.m., 11 a.m., 12 p.m., and 1 p.m.

The two cities have provided proposed building sites and building dimensions.

Projects included in this module

- **Project 1:** Baltimore, Maryland
- **Project 2:** San Francisco, California
- **Project 3:** On your own

Student worksheets

Project 1: Baltimore, Maryland

- File name: 3a_volume_shadow_worksheet.docx
- Location: EsriPress\MSDLidar\student\1_modules1_5\3volume_shadow\project1_documents

Project 2: San Francisco, California

- File name: 3b_volume_shadow_worksheet.docx
- Location: EsriPress\MSDLidar\student\1_modules1_5\3volume_shadow\project2_documents

Objectives

- Calculate the desired volume of excavation for underground parking facilities.
- Create a shadow/no-shadow map using the Skyline tool.

MAKING
SPATIAL
DECISIONS
USING GIS
AND LIDAR

1

*VOLUMETRIC
ANALYSIS AND
SHADOW MAPS*

PROJECT 1

Baltimore, Maryland

Recommended deliverables

- **Deliverable 1:** A basemap showing the location of the buildings designated for parking analysis.
- **Deliverable 2:** A volumetric analysis for the proposed parking sites.
- **Deliverable 3:** A shadow map for the proposed park area.

The questions asked in this project are both quantitative and qualitative. They identify key points that should be addressed in your analysis and final presentation.

Document your work, set environments, and examine the data

1. Open baltimore_basics1.mxd. Save the file as **balt_volume1.mxd** to the baltimore_results folder.

2. Edit the map document properties as needed for this project.

Use the Baltimore, Maryland, lidar dataset to perform volumetric analysis and produce shadow maps of the two proposed areas.

Analyze

To calculate the volume of material to be excavated, first isolate the area you must analyze. Then use the lidar data to calculate the volume of material within that area, given the topography. To create the shadow maps, convert the building footprints into 3D features based on their height. Then calculate the extent of the shadows cast by the buildings at different times. Finally, investigate the spread of the shadows on an open plaza.

DATA (Current Workspace)   \student\1_modules1_5\baltimore_data
RESULTS (Scratch Workspace)   \student\1_modules1_5\baltimore_results

59

Create a basemap of the proposed parking garages

1. Remove the City of Baltimore, water, and Baltimore lidar distribution layers.

2. Add bldgs, bldg9750, and bldg9755.

3. Make bldg9750 and bldg9755 hollow, with a red outline of 3 points. Label these two buildings by name.

4. Create a layout using appropriate cartographic principles. Save the map document as **balt_volume1.mxd**. Save the map document again as **balt_volume2.mxd**.

Deliverable 1: A basemap showing the location of the buildings designated for parking analysis.

Calculate the surface volume for the two proposed parking garages

The city wants to build two buildings (building 9750 and building 9755) for additional city government facilities. City planners have proposed a three-tier underground parking garage, which is estimated to require 30 feet of excavation. The construction company is asking for an estimated amount of excavated dirt required for each building. The Surface Volume tool can calculate the area and volume of a terrain dataset surface above or below a given plane. The following diagrams and explanations illustrate the process. A chart is provided to record analytical findings.

1. Open balt_volume2.mxd and remove the basemap. Leave building 9750 and building 9755.

2. Add baltimore_DEM from the baltresults.gdb.

3. Search for and open the Extract by Mask tool.

The Extract by Mask tool extracts the cells of a raster that correspond to the areas defined by a mask. The mask in this exercise is building 9750.

4. Enter the following settings:
   - Set the input raster to baltimore_DEM.
   - Set the input raster or feature mask data to bldg9750.
   - Set the ouput raster to baltresults.gdb**bldg9750_DEM.**

MAKING
SPATIAL
DECISIONS
USING GIS
AND LIDAR

VOLUMETRIC
ANALYSIS AND
SHADOW MAPS

5. Repeat steps 3 and 4 for building 9755. Name the file **bldg9755_DEM**.

6. Remove baltimore_DEM.

Q1 **What are the lowest and highest elevations for bldg9750_DEM?**

Q2 **What are the lowest and highest elevations for bldg9755_DEM?**

The next part of the exercise uses the Surface Volume tool. This tool calculates the surface area, the projected area, and volume of a surface relative to a given reference plane. In this example, the surfaces are the extracted building rasters and the horizontal reference plane from which calculations are derived, which, in this instance, is the ground floor of the building.

7. Search for and open the Surface Volume tool. Enter the following settings:
   • Set the input surface to bldg9750_DEM.
   • Set the output text file to baltimore_results**bldg9750_volume.**
   • Set the reference plane to ABOVE.
   • Set the plane height to **43**.

bldg9750_volume

Dataset	Plane_Height	Reference	Z_Factor	Area_2D	Area_3D	Volume
▶ ..s\baltresults.gdb\bldg9750_DEM	43	ABOVE	0.999998	35700	35879.937096	1294769.34025

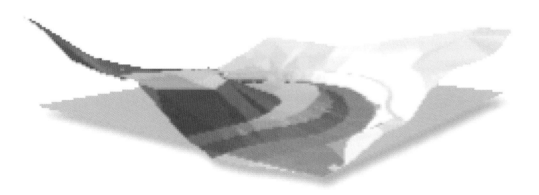

Area_2D represents the plan view area (length times width) of the building footprint. Area_3D represents the surface area. Volume represents the total volume to be evacuated between the surface and the reference height, which, in this case, represents the lowest elevation.

DATA (Current Workspace)   \student\1_modules1_5\baltimore_data
RESULTS (Scratch Workspace)   \student\1_modules1_5\baltimore_results

61

MAKING
SPATIAL
DECISIONS
USING GIS
AND LIDAR

1

VOLUMETRIC
ANALYSIS AND
SHADOW MAPS

Building 9750

89 ft

above

73 ft

If you run the Surface Volume tool with the reference plane set to ABOVE, the volume of everything from 73 to 89 feet will be calculated. To calculate the volume of the garage, set the plane height to 43 feet (73 feet – 30 feet).

To check your work, multiply the 2D area by 30 and add it to the default ABOVE calculation. The result will be a close approximation of the volume above 43 feet.

8. Repeat step 7 for building 9755.

Q3 **At what elevation should the plane height be set for building 9755?**

Q4 **Complete the following table.**

Building	Plane Height	Reference	Area_2D	Area_3D	Volume
9750					
9755					

Q5 **What factors could lead to errors in this calculation? Are you likely to underestimate or overestimate the volume of dirt to be excavated?**

9. Save the map document as **balt_volume2.mxd**.

Deliverable 2: A volumetric analysis for the proposed parking sites.

The next part of the analysis involves the use of volumetric shadow analysis to show areas of shadow and areas with no shadow in a park area. The city planners want to incorporate eating areas with benches at the designated sites. The planners have asked for a shadow map showing areas of shadows from the proposed building at 10 a.m., 11 a.m., 12 p.m., and 1 p.m.

DATA (Current Workspace)   \student\1_modules1_5\baltimore_data
RESULTS (Scratch Workspace)   \student\1_modules1_5\baltimore_results

Create a shadow map for 10 a.m., 11 a.m., 12 p.m., and 1 p.m.

1. Open balt_3D2.sxd and save the ArcScene document as **balt_shadow1.sxd**.

2. Remove all layers. Right-click Scene Layers, and go to Scene Properties. On the General tab, set the vertical exaggeration to None.

3. Update the scene document properties.

In this part of the exercise, features that have 3D properties will be exported to a multipatch feature class. Multipatch features represent the boundary of a 3D object as a single row in a database using a collection of patches. Each patch stores a range of information, including color, transparency, texture, and geometry, representing parts of a feature. You need to turn the 2D univ_plaza into a multipatch univ_plaza3D. You need the multipatch for the Intersect in 3D tool. All multipatches have z-values in the coordinates used to construct the patches.

4. Add univ_plaza. Go to Properties. Click the Extrusion tab, and then extrude height.

5. Search for and open the Layer 3D to Feature Class tool. Enter the following settings:
   • Set the input feature layer to univ_plaz.
   • Set the output feature class to baltresults.gdb\Layers**univ_plaz_3D.**

6. Remove univ_plaza and univ_plaz_3D.

7. Add left_DEM. Go to Properties. On the Base Height tab, choose the "Floating on a custom surface" option. The floating surface is left_DEM.

8. Pick an appropriate color ramp for left_DEM.

9. Add univ_plaza_bldgs and extrude by bldgsptZ.

10. Go to Scene Layers. On the General tab, calculate the vertical exaggeration.

Now you are ready to use the Sun Shadow Volume tool. This tool creates closed volumes that model shadows cast by each feature using sunlight for a given date and time.

11. Search for and open the Sun Shadow Volume tool. Enter the following settings:
    • Set the input features to univ_plaza_bldgs.
    • Set the start date and time to 8/19/2013 10:00:00 a.m. (Any date can be used. Summer and winter solstices would be good comparison dates.)
    • Click OK.

DATA (Current Workspace)   \student\1_modules1_5\baltimore_data
RESULTS (Scratch Workspace)   \student\1_modules1_5\baltimore_results

63

MAKING
SPATIAL
DECISIONS
USING GIS
AND LIDAR

1

VOLUMETRIC
ANALYSIS AND
SHADOW MAPS

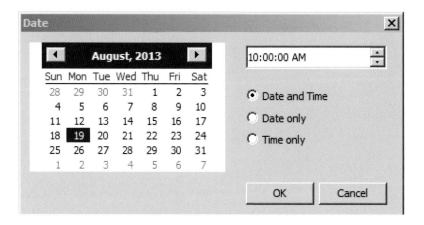

- Set the output feature class to baltresults.gdb\layers**shadows**.
- Set the end date and time to 8/19/2013 1:00 p.m.
- Set the iteration interval to 1.
- Set the iteration unit to HOURS.
- Click OK.

12. Add univ_plaza. Obtain the height from left_DEM, and give it a layer offset of **5**.

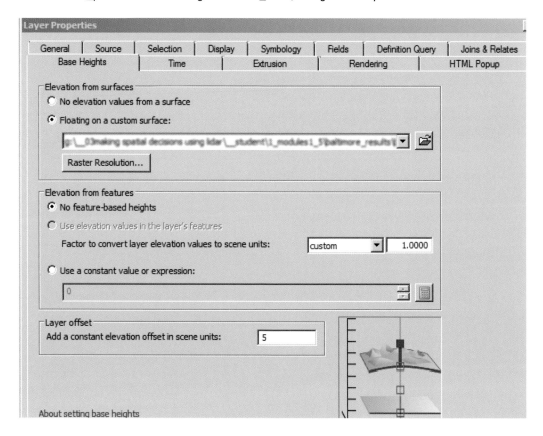

DATA (Current Workspace)   \student\1_modules1_5\baltimore_data
RESULTS (Scratch Workspace)   \student\1_modules1_5\baltimore_results

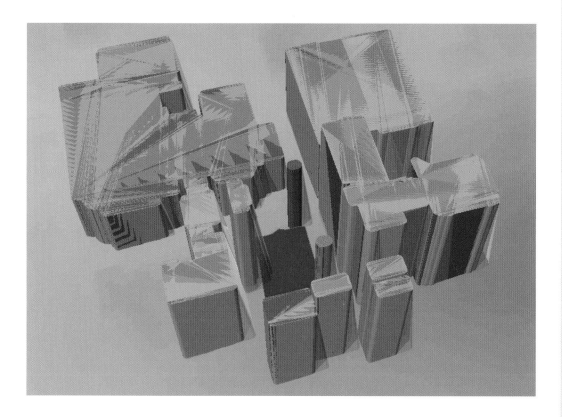

MAKING
SPATIAL
DECISIONS
USING GIS
AND LIDAR

1

VOLUMETRIC
ANALYSIS AND
SHADOW MAPS

13. Save the project as **balt_shadow1.sxd**.

The final part of the analysis is to produce a shadow map. A shadow map is a map showing areas of shadow and no shadow on the ground. For this scenario, the shadow map will show the shadows produced at hourly intervals from 10 a.m. to 1 p.m.

In addition to the shadow map, you will produce a 2D map of shadows with a basemap as a background.

14. Remove univ_plaza and add univ_plaza_3D. Obtain the height from left_DEM (Floating on a custom surface), and click Elevation from features = No feature-based heights.

15. Search for and open the Intersect 3D tool (3D Analyst). Enter the following settings:
 • Set the input multipatch features to univ_plaza_3D.
 • Set the input multipatch features to shadows.
 • Set the output feature class to baltimore_results\layers**intersect.**

DATA (Current Workspace) \student\1_modules1_5\baltimore_data
RESULTS (Scratch Workspace) \student\1_modules1_5\baltimore_results

65

16. Add intersect to the table of contents. Obtain the base height using the "Floating on a custom surface" option. The floating surface is left_DEM.
    - Select Elevation from features = No feature-based heights.
    - Give it a layer offset of **5**. Symbolize with a color.

17. Right-click intersect, and go to Properties. On the Symbology tab, on the Show panel, click Categories. Under Categories, click Unique Value. Set the value field to Date and Time.

18. Save the map document as **balt_shadow1.sxd**.

19. Go to File > Export Scene > 2D. Save the file as a PDF and name it **plaza_shadow**. Save it in the 1_module1_5\3volume_shadow\project1_documents folder.

Return to ArcMap to produce a 2D map with a basemap.

20. Open balt_volume2mxd. Save it as **balt_shadow2.mxd**.

21. Remove all files.

22. Add univ_plaza, univ_plaza_bldgs, intersect10AM, intersect11AM, intersect12PM, and inter-sect1PM. Display each intersect file with no outline and a unique color.

23. Make univ_plaza hollow with an outline width of 2. This is the defined study area and the focus for your analysis.

24. Add the Imagery with Labels basemap.

25. Create a layout using appropriate cartographic principles.

26. Save the map document as **balt_shadow2.mxd**.

27. (Optional) If you have a temporal dataset, its time properties can be set to visualize it through time using the Time Slider in ArcMap or ArcScene.
    - Right-click Intersect, and go to Properties. On the Time tab, select the "Enable time on this layer" check box.
    - Select the "Display data cumulatively" check box.
    - From the Time Slider in the main menu, activate the time series.
    - Use the playback options to slow down the playback.

DATA (Current Workspace)   \student\1_modules1_5\baltimore_data
RESULTS (Scratch Workspace)   \student\1_modules1_5\baltimore_results

Deliverable 3: A shadow map for the proposed park area.

Write a short analysis of the results of your shadow map and the viability of the park as a place for eating lunch. How would the time of year affect your analysis?

Draw conclusions and present the results

You have calculated the excavation volumes and shadow maps for new construction in Baltimore. Now choose a method of presenting your conclusions. Always keep the audience in mind as you prepare to report your results; they may not share your GIS expertise. Your results can be presented in a Word document, a PowerPoint presentation, or a more technical presentation mode, such as ArcGIS Online.

MAKING
SPATIAL
DECISIONS
USING GIS
AND LIDAR

VOLUMETRIC
ANALYSIS AND
SHADOW MAPS

DATA (Current Workspace)   \student\1_modules1_5\baltimore_data
RESULTS (Scratch Workspace)   \student\1_modules1_5\baltimore_results

67

MAKING
SPATIAL
DECISIONS
USING GIS
AND LIDAR

2

VOLUMETRIC
ANALYSIS AND
SHADOW MAPS

San Francisco, California

Recommended deliverables

- **Deliverable 1:** A basemap showing the location of the buildings designated for parking analysis.
- **Deliverable 2:** A volumetric analysis for the proposed parking sites.
- **Deliverable 3:** A shadow map for the proposed park area.

The questions asked in this project are both quantitative and qualitative. They identify key points that should be addressed in your analysis and final presentation.

Document your work, set environments, and examine the data

1. Open sanfran_basics1.mxd. Save the file as **sanfran_volume1.mxd** to the sanfrancisco_results folder.

2. Edit the map document properties as needed for this project.

Use the San Francisco, California, lidar dataset to perform volumetric analysis and produce shadow maps of two proposed areas.

DATA (Current Workspace)   \student\1_modules1_5\sanfrancisco_data
RESULTS (Scratch Workspace)   \student\1_modules1_5\sanfrancisco_results

Analyze

Create a basemap of the proposed parking garages

1. Remove sanfran.lasd and san_francisco.

2. Add bldgs, bldg100, and bldg200.

3. Create a layout using appropriate cartographic principles. Save the map document as **sanfran_volume1.mxd**. Save the map document again as **sanfran_volume2.mxd**.

Deliverable 1: A basemap showing the location of the buildings designated for parking analysis.

Calculate the surface volume for the two proposed parking garages

The city wants to build two buildings (building 100 and building 200) for additional city government facilities. City planners have proposed a three-tier underground parking garage, which is estimated to require 30 feet of evacuation. The construction company is asking for an estimated amount of excavated dirt required for each building. The San Francisco dataset has a linear unit of meters. 30 feet = 9.144 meters. Use 9 meters for the excavation depth.

1. Open sanfran_volume2.mxd and remove the basemap. Leave building 100 and building 200.

2. Add SF_DEM from the sanfrancisco.gdb.

3. Use the Extract by Mask tool to extract the rasters for the two buildings. Name the two files **bldg100_DEM** and **bldg200_DEM,** and save them to \sanfrancisco_results\sanfrancisco.gdb.

Q1   What are the lowest and highest elevations in bldg100_DEM?

Q2   What are the lowest and highest elevations in bldg200_DEM?

4. Using the Surface Volume tool, determine the volume to be excavated. Save the files as **bldg100_volume** and **bldg200_volume** to the sanfrancisco_results folder.

Q3   At what elevation should the plane height be set for bldg100?

Q4   At what elevation should the plane height be set for bldg200?

DATA (Current Workspace)   \student\1_modules1_5\sanfrancisco_data
RESULTS (Scratch Workspace)   \student\1_modules1_5\sanfrancisco_results

69

Q5 ***Complete the following table.***

Building	Plane Height	Reference	Area_2D	Area_3D	Volume
100					
200					

Remember: Check your work.

Q6 ***What factors could lead to errors in this calculation? Are you likely to underestimate or overestimate the volume of dirt to be excavated?***

5. Save the map document as **sanfran_volume2.mxd**.

Deliverable 2: A volumetric analysis for the proposed parking sites.

The next part of the analysis involves the use of volumetric shadow analysis to show areas of shadow and areas with no shadow in the Yerba Buena/Sister City Gardens area. The city planners want to incorporate eating areas with benches at the Gardens site. The planners have asked for a shadow map showing areas of shadows at 10 a.m., 11 a.m., 12 p.m., and 1 p.m.

Create a shadow map for 10 a.m., 11 a.m., 12 p.m., and 1 p.m.

1. Open SF_3D2.sxd and save the map document as **SF_shadow1.sxd**.

2. Remove all layers, set the vertical exaggeration to None, and update the scene document properties.

3. Use the Layer 3D to Feature tool. Save the file as **garden_3D** in the sanfrancisco_results folder.

4. Add SF_DEM, bldgs_shadow, and gardens_2D.

Now you are ready to use the Sun Shadow Volume tool. This tool creates closed volumes that model shadows cast by each feature using sunlight for a given date and time.

5. Use the Sun Shadow Volume tool to calculate the shadows for 10 a.m., 11 a.m., 12 p.m., and 1 p.m. on August 22, 2013. (Any date can be used. Consider using times of the year when shadows are the longest.) Set the output feature class to sanfrancisco.gdb\layers\shadows.

6. Save the project as **SF_shadow1.sxd**.

DATA (Current Workspace)   \student\1_modules1_5\sanfrancisco_data
RESULTS (Scratch Workspace)   \student\1_modules1_5\sanfrancisco_results

The final part of the analysis is to produce a shadow map. A shadow map is a map showing areas of shadow and no shadow on the ground. For this scenario, the shadow map will show the shadows cast at hourly intervals from 10 a.m. to 1 p.m.

In addition to the shadow map, you will produce a 2D map of shadows with a basemap as a background.

7. Add gardens_3D. Use the Intersect 3D tool (3D Analyst) to create an intersection of the park and shadow for each time. Save the files in the sanfrancisco_results folder.

8. Save the map document as **SF_shadow2.sxd**.

9. Open SF_volume2.mxd. Save it as **balt_shadow2.mxd**.

10. Add the intersect and the Imagery with Labels basemap.

11. Create a layout using appropriate cartographic principles.

12. Save the map document as **balt_shadow2.mxd**.

Optional: Time-enable your map.

Deliverable 3: A shadow map for the proposed park area.

Write a short analysis of the results of your shadow map and the viability of the park as a place for eating lunch. How would the time of year affect your analysis?

MAKING
SPATIAL
DECISIONS
USING GIS
AND LIDAR

VOLUMETRIC
ANALYSIS AND
SHADOW MAPS

Draw conclusions and present the results

You have calculated the excavation volumes and shadow maps for San Francisco. Now choose a method of presenting your conclusions. Always keep the audience in mind as you prepare to report your results; they may not share your GIS expertise. Your results can be presented in a Word document, a PowerPoint presentation, or a more technical presentation mode, such as ArcGIS Online.

DATA (Current Workspace)   \student\1_modules1_5\sanfrancisco_data
RESULTS (Scratch Workspace)   \student\1_modules1_5\sanfrancisco_results

71

MAKING
SPATIAL
DECISIONS
USING GIS
AND LIDAR

3

VOLUMETRIC
ANALYSIS AND
SHADOW MAPS

PROJECT 3

On your own

Scenario

You have worked through a guided project and repeated the analysis for another project. In this project, you will reinforce your skills by researching and analyzing a similar scenario entirely on your own. First, identify your study area and acquire data for your analysis. You may want to study a local area.

Here is a list of topics that have been studied using lidar. The following represent possible project ideas:

- Forest characterization—canopy height and density
- Flood modeling
- Finding faults
- Geomorphic mapping
- Stream slope
- Archaeology field campaigns
- Mining—calculation of ore volumes
- Wind farm optimization

Many different websites distribute lidar data. Most of the available lidar data is within the United States. There are national data sites, and there are state data distributors. Reading the metadata and identifying a specific area of study is critical before downloading files. Some lidar files come compressed and require third-party software to convert them into the LAS format used by Esri.

- Open Topography facilitates community access to high-resolution, Earth science-oriented, topography data. It is a National Science Foundation-funded data facility. http://www.opentopography.org.
- The Earth Explorer provides online access to remotely sensed data from the US Geological Survey Earth Resources Observation and Science (EROS) Center archive. http://earthexplorer.usgs.gov/.
- National Oceanic and Atmospheric Administration, a world leader in coast science and management, provides state lidar datasets. http://www.csc.noaa.gov/dataviewer/#.
- Wikipedia lists national lidar datasets, organized according to state. http://en.wikipedia.org/wiki/National_LIDAR_Dataset_%E2%80%93_USA.

Research

Research the problem and answer the following questions:
1. What is the area of study?
2. What problem are you going to study?
3. What data is available?

VISIBILITY ANALYSIS COMPARISON

SCENARIO ···

Baltimore and San Francisco both want to provide their residents with continuous cell phone coverage. A new telecommunications company has approached the two cities with a plan to do a visibility analysis of the designated geographic area to enhance coverage. Visibility analysis shows how topography affects cell phone coverage. However, the placement of cell phone towers (structures that broadcast cell phone signals) is a complicated process that also requires additional sophisticated models. This analysis uses viewsheds and a line of sight to determine what areas are visible from particular cell phone towers. Telecommunication companies want to find the locations for their cell phone towers that provide the maximum signal coverage in a geographic area. Viewsheds are usually generated using digital elevation models (DEMs). However, digital surface models (DSMs) generated from lidar have greater accuracy because they include surface features, such as buildings and trees.

In this analysis, the viewshed will be determined using preexisting cell phone towers and a DSM. After the initial viewshed analysis, potential problem areas will be examined using line-of-sight analysis.

Projects included in this module

- **Project 1:** Baltimore, Maryland
- **Project 2:** San Francisco, California
- **Project 3:** On your own

Student worksheets

Project 1: Baltimore, Maryland

- File name: 4a_cellular_network_worksheet.docx
- Location: EsriPress\MSDLidar\student\1_modules1_5\4cellular_network\project1_documents

Project 2: San Francisco, California

- File name: 4b_cellular_network_worksheet.docx
- Location: EsriPress\MSDLidar\student\1_modules1_5\4cellular_network\project2_documents

Objectives

- Calculate a viewshed for the designated study area using preexisting cell phone towers and lidar DSM.
- Identify areas of least visibility. Perform additional line-of-sight analysis from existing cell phone towers and randomly generated points.
- Visualize the line-of-sight analysis in 3D.

PROJECT ①

Baltimore, Maryland

Recommended deliverables

- **Deliverable 1:** A basemap of the study area, DSM, and existing cell phone towers.
- **Deliverable 2:** A map showing the result of your visibility calculation using a viewshed, lidar, and existing cell phone towers.
- **Deliverable 3:** A map showing the results of your analysis using line of sight, cell phone towers, and randomly generated points.

The questions asked in this project are both quantitative and qualitative. They identify key points that should be addressed in your analysis and final presentation.

Document your work, set environments, and examine the data

1. Open ArcMap. Edit the map document properties as needed for this project.

2. Set environments:
   - Open the data frame properties. Set the map projection to Projected Coordinate Systems > State Plane > NAD_1983 > StatePlane_Maryland_FIPS_1900_Feet.
   - On the Geoprocessing menu, click Environments.
   - Expand Workspace, and set the current workspace to \student\1_modules1_5\baltimore_data.
   - Set the scratch workspace to \student\1_modules1_5\baltimore_results.
   - For the output coordinate system, select Same as Display.

3. Save the map document as **balt_cell1** in the baltimore_results folder.

DATA (Current Workspace)   \student\1_modules1_5\baltimore_data
RESULTS (Scratch Workspace)   \student\1_modules1_5\baltimore_results

77

MAKING
SPATIAL
DECISIONS
USING GIS
AND LIDAR

VISIBILITY
ANALYSIS
COMPARISON

Analyze

Create a basemap of the study area and cell phone towers

1. Add baltimore_city, towers, and water from the baltimore.gdb.

2. Add baltimore_tiles.lasd from the baltimore_results folder.

3. Symbolize the towers and add an appropriate basemap. The towers are categorized as follows:
   - STRUCTYPE = 1 tower, 200 feet high
   - STRUCTYPE = 2 buildings, height of the buildings plus 100-foot antennae

Save the symbolization you created as a layer file.

4. Right-click Structures, and go to Save as Layer File. Name the file **Structures**, and save it to the baltimore_results folder.

5. Save the file as **balt_cell1** to the baltimore_results folder. Save it again as **balt_cell2** to the baltimore_results folder.

Deliverable 1: A basemap of the study area, DSM, and existing cell phone towers.

Perform a visibility calculation using lidar DSM and existing cell phone towers

1. Open the balt_cell2 map document and remove baltimore_city, water, and online basemap.

2. Add baltimore_DSM from the baltresults.gdb.

Q1　　*Describe the landscape of the study area in Baltimore city. What structures are visible? Where are the tallest buildings? Adding the imagery basemap with labels will help you identify structures.*

Q2　　*What is the spatial distribution of the cell phone towers?*

Q3　　*Why are there no surface values for the water area?*

DATA (Current Workspace)　\student\1_modules1_5\baltimore_data
RESULTS (Scratch Workspace)　\student\1_modules1_5\baltimore_results

A viewshed identifies the cells in an input raster that are visible from one or more observation locations. In this exercise, the lidar DSM is the input raster. Remember, the DSM uses the lidar first return, which includes buildings and tree canopy. Buildings are clearly visible in the baltimore_DSM. The cell phone towers will be used as the observation locations. All cells that cannot be "seen" from the observer points are given a value of 0 (not visible) and the cells that can be "seen" are given a value of 1 (visible).

The towers (observation points) must be given a height value. Remember, STRUCTYPE 1 is a 200 foot-high tower, and STRUCTYPE 2 is a building, the height of which is added to an antennae height of 100 feet.

The visibility analysis can be controlled by designating the vertical height to be added to the z-value of a location on the surface. This is designated as OFFSETA and will correspond to the tower height. The vertical height added to the z-value of each cell considered for visibility is designated as OFFSETB. For example, OFFSETB might correspond to the height of the eyes of an observer on the ground (typically 5–6 feet).

3. Right-click towers, and go to Data > Export Data. Set the output feature class to baltimore_results\baltresults.gdb\layers**towers2**.

4. Remove towers. Add the Structures.lyr file to towers2.

5. Open the towers2 attribute table. From the Table Options menu in the upper left corner, click Add field. Name the new field **OFFSETA,** and set the type to Short Integer. Click OK.

6. Select by attributes. Set the STRUCTYPE = 1. Right-click the OFFSETA field, and click Field Calculator. In the Field Calculator, enter **200**.

7. Repeat step 6, but set the STRUCTYPE = 2. In the Field Calculator, replace 200 with **100**.

8. Clear all selections.

9. Add another field, and name it **OFFSETB**. Set the type to Short Integer.

10. Right-click the OFFSETB field, and click Field Calculator. In the Field Calculator, enter **5**.

MAKING
SPATIAL
DECISIONS
USING GIS
AND LIDAR

*VISIBILITY
ANALYSIS
COMPARISON*

DATA (Current Workspace) \student\1_modules1_5\baltimore_data
RESULTS (Scratch Workspace) \student\1_modules1_5\baltimore_results

79

MAKING
SPATIAL
DECISIONS
USING GIS
AND LIDAR

VISIBILITY
ANALYSIS
COMPARISON

11. Search for and open the Viewshed tool (Spatial Analyst). Enter the following settings:
- Set the input raster to baltimore_DSM.
- Set the input point observer features to towers (Structures).
- Set the output raster to baltimore_results\baltresults.gdb**viewshed**.
- Click OK.

Q4 *Which quadrant has the best visibility? Which quadrant has the worst visibility?*

Q5 *What is the relationship between the number of cell phone towers and visibility?*

12. Search for and open the Hillshade tool (Spatial Analyst). Enter the following settings:
- Set the input raster to baltimore_DSM.
- Set the output raster to the baltimore_results\baltresults.gdb**balt_shade**.
- Accept the default hillshade azimuth and altitude settings.

13. Move the viewshed above balt_shade. In the viewshed display properties, set the transparency to **50%**.

14. Remove baltimore_DSM.

15. Save the map document as **balt_cell2**.

Deliverable 2: A map showing the result of your visibility calculation using a viewshed, lidar, and existing cell phone towers.

Create another map using line of sight and cell phone towers

After looking at the initial viewshed map, the City of Baltimore requested a more detailed analysis of the southeast quadrant. Specifically, the City requested a line-of-sight analysis between the existing cell phone towers and random points.

The Line Of Sight tool calculates visibility between an observer and a target given its position in 3D space relative to the obstructions provided by the surface (DSM). In this exercise, the observer points will be the cell phone towers. An observer offset can be used to account for the height of the cell phone towers, and a target offset can be used to account for the height of the target. In this exercise, random points will be used as targets with a vertical offset of 5 feet.

DATA (Current Workspace)   \student\1_modules1_5\baltimore_data
RESULTS (Scratch Workspace)   \student\1_modules1_5\baltimore_results

1. Open ArcMap. Edit the map document properties as needed for this project.

2. Set environments:
   - Open the data frame properties. Set the map projection to Projected Coordinate Systems > State Plane > NAD_1983 > StatePlane_Maryland_FIPS_1900_Feet.
   - On the Geoprocessing menu, click Environments.
   - Expand Workspace, and set the current workspace to \student\1_modules1_5\ baltimore_data.
   - Set the scratch workspace to \student\1_modules1_5\baltimore_results.
   - For the output coordinate system, select Same as Display.

3. Save the map document as **balt_cell3** to the baltimore_results folder.

4. Add sa_cell and water from the baltimore_data\baltimore.gdb.

5. Add towers2 and baltimore_DSM from the baltimore_results\baltresults.gdb.

6. Search for and open the Clip tool (Analysis). Enter the following settings:
   - Set the input feature to towers2.
   - Set the clip feature to sa_cell.
   - Set the output feature class to baltimore_results\baltresults.gdb\layers**SE_towers.**
   - Click OK.

7. Remove towers. Zoom to the SE_towers data layer.

8. Repeat the process previously described, but clip water. Call the output file **SE_water,** and save it to baltresults.gdb\layers. Remove water.

9. Search for and open the Extract by Mask tool (Spatial Analyst). Enter the following settings:
   - Set the input raster to baltimore_DSM.
   - Set the input raster or feature mask data to sa_cell.
   - Set the output raster to baltimore_results\balt_results.gdb**SE_DSM.**

10. Remove baltimore_DSM.

11. Classify SE_ towers.

MAKING
SPATIAL
DECISIONS
USING GIS
AND LIDAR

VISIBILITY
ANALYSIS
COMPARISON

DATA (Current Workspace)   \student\1_modules1_5\baltimore_data
RESULTS (Scratch Workspace)   \student\1_modules1_5\baltimore_results

81

MAKING
SPATIAL
DECISIONS
USING GIS
AND LIDAR

1

VISIBILITY
ANALYSIS
COMPARISON

12. Search for and open the Create Random Points tool (Data Management). Enter the following settings:
- Set the output location to baltimore_results\baltresults.gdb\layers.
- Set the output point feature class to **random_pts**.
- Set the constraining extent to sa_cell.
- Set the number of points to **20.**
- Click OK.

13. If any of the random points are in the water and need to be deleted, open the Editor toolbar, start editing, and edit random_pts. Select the points that were generated in the water and delete them. Stop and save your edits.

14. Open the random_pts attribute table. Add a short integer field named **OFFSETB**. Populate the field with the value **5**.

DATA (Current Workspace)   \student\1_modules1_5\baltimore_data
RESULTS (Scratch Workspace)   \student\1_modules1_5\baltimore_results

Creating a line of sight is a two-part process and must be done individually for each observer point. In this exercise, you will perform a line-of-sight analysis with cell phone towers 1, 3, 4, 7, and 8.

15. Select SE_tower with OBJECTID 1.

16. Search for and open the Construct Sight Lines tool (3D Analyst). Enter the following settings:
    • Set the observer points to SE_towers.
    • Set the target features to random_pts.
    • The height field that the software is looking for is the observer height OFFSETA and the target height OFFSETB.
    • Set the output to baltresults.gdb\layers**lineofsight1.**
    • Click OK.

17. Repeat steps 15–16 for towers 3, 4, 7, and 8. Select the SE_Tower with each OBJECTID value. Name the outputs appropriately.

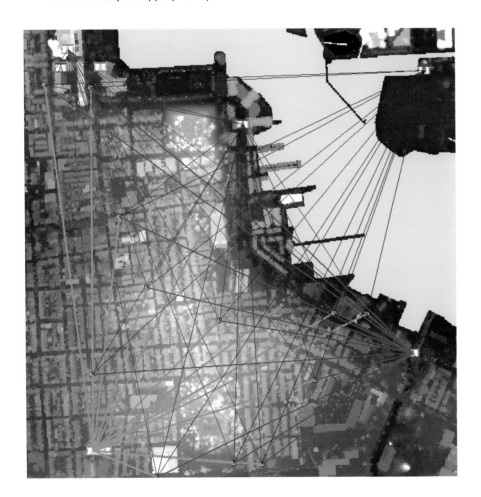

MAKING
SPATIAL
DECISIONS
USING GIS
AND LIDAR

1

VISIBILITY
ANALYSIS
COMPARISON

DATA (Current Workspace)   \student\1_modules1_5\baltimore_data
RESULTS (Scratch Workspace)   \student\1_modules1_5\baltimore_results

83

MAKING
SPATIAL
DECISIONS
USING GIS
AND LIDAR

1

VISIBILITY
ANALYSIS
COMPARISON

The constructed lines of sight will now be used to determine the visibility of sight lines over obstructions represented in the SE_DSM.

18. Search for and open the Line Of Sight tool. Enter the following settings:
    - Set the input surface to SE_DSM.
    - Set the input line features to linesofsight1.
    - Set the output feature class to baltresults.gdb\layers**LOS_1**.

19. Repeat step 18 for lineofsight3, lineofsight4, lineofsight7, and lineofsight8. Name the output files appropriately.

20. Remove the sight lines (lineofsight).

21. Symbolize the visible part of the line with a larger line width.

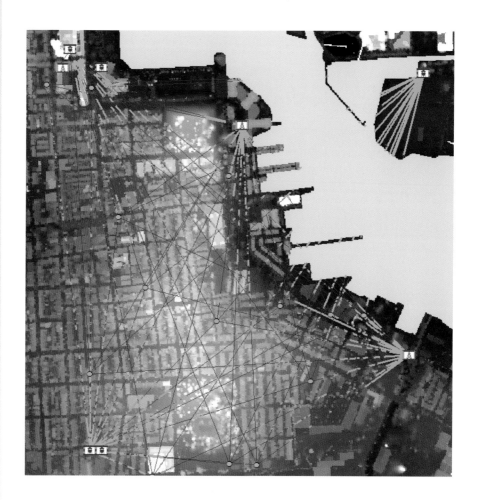

DATA (Current Workspace)   \student\1_modules1_5\baltimore_data
RESULTS (Scratch Workspace)   \student\1_modules1_5\baltimore_results

MAKING
SPATIAL
DECISIONS
USING GIS
AND LIDAR

VISIBILITY
ANALYSIS
COMPARISON

22. Search for and open the Hillshade tool. Enter the following settings:
- Set the input raster to SE_DSM.
- Set the output raster to baltimore_results\baltresults.gdb**SA_SE_Shade**.

23. Remove SE_DSM.

Q6 ***What section of the study area has the least visibility from the lines of sight shown? Why?***

Q7 ***How could visibility be improved?***

Q8 ***What factor in our analysis could be changed to provide a better analysis?***

24. Save the map as **balt_cell3**.

Deliverable 3: A map showing the results of your analysis using line of sight, cell phone towers, and randomly generated points.

Draw conclusions and present the results

Now that you have mapped the visibility for the cell phone towers in Baltimore and completed a line-of-sight analysis, choose a method of presenting your conclusions. Always keep the audience in mind as you prepare to report your results; they may not share your GIS expertise. Your results can be presented in a Word document, a PowerPoint presentation, or a more technical presentation mode, such as ArcGIS Online.

DATA (Current Workspace)   \student\1_modules1_5\baltimore_data
RESULTS (Scratch Workspace)   \student\1_modules1_5\baltimore_results

85

MAKING
SPATIAL
DECISIONS
USING GIS
AND LIDAR

2

*VISIBILITY
ANALYSIS
COMPARISON*

PROJECT **2**

San Francisco, California

Lidar data was obtained from the US Geological Survey Earth Explorer website: http://earthexplorer.usgs.gov/.

Building footprints were obtained from the SF OpenData website: https://data.sfgov.org/Other/Data-Catalog/h4ui-ubbu.

▍ Recommended deliverables

- **Deliverable 1:** A basemap of the study area, DSM, and existing cell phone towers.
- **Deliverable 2:** A map showing the result of your visibility calculation using a viewshed, lidar, and existing cell phone towers.
- **Deliverable 3:** A map showing the results of your analysis using line of sight, cell phone towers, and randomly generated points.

The questions asked in this project are both quantitative and qualitative. They identify key points that should be addressed in your analysis and final presentation.

▍ Document your work, set environments, and examine the data

1. Open ArcMap. Edit the map document properties as needed for this project.

2. Set environments:
    - Open the data frame properties. Set the map projection to Projected Coordinate Systems > UTM > NAD_1983 > Zone_10N.
    - On the Geoprocessing menu, click Environments.
    - Expand Workspace, and set the current workspace to \student\1_modules1_5\ sanfrancisco_data.

DATA (Current Workspace)   \student\1_modules1_5\sanfrancisco_data
RESULTS (Scratch Workspace)   \student\1_modules1_5\sanfrancisco_results

- Set the scratch workspace to \student\1_modules1_5\sanfrancisco_results.
- For the output coordinate system, select Same as Display.

Analyze

MAKING
SPATIAL
DECISIONS
USING GIS
AND LIDAR

VISIBILITY
ANALYSIS
COMPARISON

Create a basemap of the study area and cell phone towers

1. Add sa_sanfrancisco and towers from sanfrancisco_data\sanfrancisco.gdb\layers.

2. Add sanfran.lasd from the sanfrancisco_results folder.

3. Symbolize the towers and add an appropriate background map. The towers are categorized as follows:
   - STRUCTYPE = 1 tower, 50 feet high
   - STRUCTYPE = 2 buildings, height of the building plus a 10-foot antennae

4. Name the file **SF_cell1**, and save it to the sanfrancisco_results folder. Save the file again as **SF_cell2** to the sanfrancisco_results folder.

Deliverable 1: A basemap of the study area, DSM, and existing cell phone towers.

Perform a visibility calculation using lidar DSM and existing cell phone towers

1. Open SF_cell2. Remove sanfran.lasd and the background map. Add SF_DSM from sanfrancisco_results\sanfrancisco.gdb.

Q1   *Describe the landscape of the study area in the San Francisco area. What structures are visible? Where are the tallest buildings?*

Q2   *What is the spatial distribution of the cell phone towers?*

Q3   *Why are there no surface values for the water area?*

2. Complete a viewshed analysis of the San Francisco area.

3. Display the viewshed visibility analysis on a hillshade made from SF_DSM.

Q4   *Which quadrant has the best visibility? Which quadrant has the worst visibility?*

DATA (Current Workspace)   \student\1_modules1_5\sanfrancisco_data
RESULTS (Scratch Workspace)   \student\1_modules1_5\sanfrancisco_results

87

MAKING
SPATIAL
DECISIONS
USING GIS
AND LIDAR

VISIBILITY
ANALYSIS
COMPARISON

Q5 *What is the relationship between the number of cell phone towers and visibility?*

4. Save the map document as **SF_cell2**.

Deliverable 2: A map showing the result of your visibility calculation using a viewshed, lidar, and existing cell phone towers.

Create another map using line of sight and cell phone towers

Citizens have asked for another line-of-sight analysis for a particular area that contains South Park and AT&T Park. This area has been designated as another study area.

1. Add sa_cell from the sanfrancisco.gdb.

2. Generate 20 random points and add 5 feet of height. Name the field **OFFSETB**.

3. Perform a line-of-sight analysis.

4. Display on a hillshade of San Francisco DSM.

Q6 *What section of the study area has the least visibility from the lines of sight shown? Why?*

Q7 *What factor in our analysis could be changed to provide a better analysis?*

Deliverable 3: A map showing the results of your analysis using line of sight, cell phone towers, and randomly generated points.

▌ Draw conclusions and present the results

Now that you have mapped the visibility for the cell phone towers in San Francisco and completed a line-of-sight analysis, choose a method of presenting your conclusions. Always keep the audience in mind as you prepare to report your results; they may not share your GIS expertise. Your results can be presented in a Word document, a PowerPoint presentation, or a more technical presentation mode, such as ArcGIS Online.

DATA (Current Workspace)   \student\1_modules1_5\sanfrancisco_data
RESULTS (Scratch Workspace)   \student\1_modules1_5\sanfrancisco_results

PROJECT **3**

On your own

Scenario

You have worked through a guided project and repeated the analysis for another project. In this project, you will reinforce your skills by researching and analyzing a similar scenario entirely on your own. First, identify your study area and acquire data for your analysis. You may want to study a local area.

Here is a list of topics that have been studied using lidar. The following represent possible project ideas:

- Forest characterization—canopy height and density
- Flood modeling
- Finding faults
- Geomorphic mapping
- Stream slope
- Archaeology field campaigns
- Mining—calculation of ore volumes
- Wind farm optimization

MAKING
SPATIAL
DECISIONS
USING GIS
AND LIDAR

3

VISIBILITY
ANALYSIS
COMPARISON

Many different websites distribute lidar data. Most of the available lidar data are within the United States. There are national data sites, and there are state data distributors. Reading the metadata and identifying a specific area of study is critical before downloading files. Some lidar files come compressed and require third-party software to convert them into the LAS format used by Esri.

- Open Topography facilitates community access to high-resolution, Earth science-oriented, topography data. It is a National Science Foundation-funded data facility. http://www.opentopography.org.
- The Earth Explorer provides online access to remotely sensed data from the US Geological Survey Earth Resources Observation and Science (EROS) Center archive. http://earthexplorer.usgs.gov/.
- National Oceanic and Atmospheric Administration, a world leader in coast science and management, provides state lidar datasets. http://www.csc.noaa.gov/dataviewer/#.
- Wikipedia lists national lidar datasets, organized according to state. http://en.wikipedia.org/wiki/National_LIDAR_Dataset_%E2%80%93_USA.

Research

Research the problem and answer the following questions:

1. What is the area of study?
2. What problem are you going to study?
3. What data is available?

SURGING SEAS

SCENARIO

As your final contractual obligation to Baltimore and San Francisco, the cities want you to construct new flood insurance rate maps (FIRMs) for the Federal Emergency Management Agency (FEMA). FEMA knows that these new lidar-based FIRMs will be much more accurate than their older maps. These FIRM maps are used to establish flood insurance rate premiums, so their accuracy is critical. FEMA requests that the maps be published on ArcGIS Online so that citizens can use their address to access the map information. FEMA wants the FIRM maps to be constructed in line with the following hurricane inundation zones:

- 5 feet above normal
- 8 feet above normal
- 12 feet above normal
- 18 feet+ above normal

Projects included in this module

- **Project 1:** Baltimore, Maryland
- **Project 2:** San Francisco, California
- **Project 3:** On your own

Student worksheets

Project 1: Baltimore, Maryland

- File name: 5a_surge_worksheet.docx
- Location: EsriPress\MSDLidar\student\1_modules1_5\5surge\project1_documents

Project 2: San Francisco, California

- File name: 5b_surge_worksheet.docx
- Location: EsriPress\MSDLidar\student\1_modules1_5\5surge\project2_documents

Objectives

- Create raster elevation datasets using the LAS to Raster tool.
- Use the Raster Calculator tool to designate flood zones for 30-meter DEMs (digital elevation models) and lidar rasters.
- Calculate and compare differences between DEM flood zones and lidar-calculated flood zones.

A 30-meter DEM may be downloaded from the US Geological Survey National Map Viewer website: http://viewer.nationalmap.gov/viewer/.

MAKING
SPATIAL
DECISIONS
USING GIS
AND LIDAR

1

SURGING
SEAS

PROJECT 1

Baltimore, Maryland

Recommended deliverables

- **Deliverable 1:** A map comparing traditional 30-meter elevation data to lidar that also includes the number of buildings affected by the various storm surge delineations.

The questions asked in this project are both quantitative and qualitative. They identify key points that should be addressed in your analysis and final presentation.

Document your work, set environments, and examine the data

1. Open baltimore_basics1.mxd. Save the file as **balt_elevation1.mxd** to the baltimore_results folder.

2. Edit the map document properties as needed for this project.

This map document shows the construction of new FIRMs for FEMA.

Analyze

Convert data to the correct projection and linear unit

1. Remove all layers.

2. Add baltimore_DEM, which is the raster that you made from the LAS dataset.

You know that the map projection is NAD_1983_StatePlane_Maryland_FIPS_1900_Feet. You also know that the linear measurement is in feet.

DATA (Current Workspace)   \student\1_modules1_5\baltimore_data
RESULTS (Scratch Workspace)   \student\1_modules1_5\baltimore_results

MAKING
SPATIAL
DECISIONS
USING GIS
AND LIDAR

1

*SURGING
SEAS*

3. Zoom to a pixel and measure its spatial dimensions. Alternatively, you can access this information by right-clicking the pixel, going to baltimore_DEM > Properties > Source, and then noting the cell size.

Q1   ***What is the spatial dimension of one pixel of baltimore_DEM?***

4. Add balt_elev_NED. Right-click balt_elev_NED, and go to Data > View Item Description.

Q2   ***What is the map projection and the linear unit?***

To do a meaningful analysis, both of these datasets must be in the same map projection and have the same linear unit.

Q3   ***What is the spatial dimension of one pixel of baltimore_elev_MD_ft?***

5. In the table of contents, select baltimore_DEM. Rename it **LIDAR and balt_elev_MD_ft to NED**. NED stands for national elevation data.

Q4   ***How do the elevation ranges differ in the two files?***

Use the Raster Calculator tool to establish surge areas

You are now ready to start making the FIRM maps. FEMA has established the following hurricane inundation zones. Use these values to make your FIRM maps:

- Category I Hurricane   Storm Surge 0–5 ft
- Category II Hurricane   Storm Surge 8 ft
- Category III Hurricane   Storm Surge 12 ft
- Category IV Hurricane   Storm Surge 18 ft

1. Search for and open the Raster Calculator tool to create the Category I hurricane inundation zone. Use the following Map Algebra expression. It is important that you get the syntax correct. ("NED" >= 0) & ("NED" <= 5)

2. Name the output raster **NED_5** and save it to baltimore_results\baltresults.gdb.

3. Repeat step 1 for LIDAR and name the output raster **LIDAR_5**.

DATA (Current Workspace)   \student\1_modules1_5\baltimore_data
RESULTS (Scratch Workspace)   \student\1_modules1_5\baltimore_results

95

MAKING
SPATIAL
DECISIONS
USING GIS
AND LIDAR

1

SURGING
SEAS

You have executed a Map Algebra expression where all areas that fit your criteria of elevations between 0–5 feet are assigned a value of 1, and other areas are assigned a value of 0.

4. Symbolize the 0 values with no color. Symbolize the 1 values with different colors for NED_5 and LIDAR_5 to help you distinguish between the NED and the LIDAR flooded areas.

5. Name the data frame **Storm Surge 0–5 ft**.

6. Insert a new data frame and name it **Storm Surge 8 ft and Below**. Copy LIDAR and NED into the new data frame.

7. Repeat steps 1 and 2, but use the following Map Algebra expression: ("NED" >= 0) & ("NED" <= 8).

8. Name the files **NED_8** and **LIDAR_8**, and save them to the output rasters in baltimore_results\baltresults.gdb.

9. Symbolize the 0 values with no color. Symbolize the 1 values with different colors for NED_8 and LIDAR_8 to help you distinguish between the NED and the LIDAR flooded areas.

10. Insert a new data frame and name it **Storm Surge 12 ft and Below**. Copy LIDAR and NED into the new data frame.

11. Repeat steps 1 and 2, using the appropriate Map Algebra expression.

12. Name the output rasters **NED_12** and **LIDAR_12**, and save them to baltimore_results\baltresults.gdb. Symbolize each in a similar manner as before.

13. Insert a new data frame and name it **Storm Surge 18 ft and Below**. Copy LIDAR and NED into the new data frame.

14. Repeat steps 1 and 2, but use the appropriate Map Algebra expression.

15. Name the output rasters **NED_18** and **LIDAR_18**, and save them to baltimore_results \baltresults.gdb. Symbolize each as before.

Q5 ***Look at the four data frames and compare the lidar and NED results. Why are they different?***

DATA (Current Workspace) \student\1_modules1_5\baltimore_data
RESULTS (Scratch Workspace) \student\1_modules1_5\baltimore_results

Convert the flood surge rasters into polygons

MAKING
SPATIAL
DECISIONS
USING GIS
AND LIDAR

1

SURGING
SEAS

To analyze how many buildings are affected by the storm surge, convert the Boolean rasters you created previously into polygons. When you convert the rasters to polygons, the grid code numbers will, of course, be either 1 or 0. You want to isolate the portion of the polygon that has a grid code of 1, which represents the flooded area.

1. Activate the Storm Surge 0–5 ft data frame.

2. Search for and open the Raster to Polygon tool. Enter the following settings:
   - Set the input raster to NED_5.
   - Set the field to Value.
   - Set the output polygon features to \baltimore_results\baltresults.gdb\Layers**NED5poly.**

3. For the new NED5poly feature class, use the Select By Attributes tool. Enter gridcode = 1.

4. Export the selected features by right-clicking NED5poly, and going to Data > Export Data. Name the feature class **NED5flood**, and save it to the baltresults.gdb\layers dataset.

5. Remove NED5poly.

6. Repeat steps 2–5 with LIDAR_5. Name the final feature class **LIDAR5flood**.

7. Activate the Storm Surge 8 ft and Below data frame, and repeat steps 1–6. Name the final feature classes **NED8flood** and **LIDAR8flood**.

8. Activate the Storm Surge 12 and Below data frame, and repeat steps 1–6. Name the final feature classes **NED12flood** and **LIDAR12flood**.

9. Activate the Storm Surge 18 and Below data frame, and repeat steps 1–6. Name the final feature classes **NED18flood** and **LIDAR18flood**.

Intersect the buildings with the storm surge polygons

There is a significant difference in the number of buildings that are affected when the lidar data is used to make the flood map rather than the NED data.

1. Activate the Storm Surge 0–5 ft data frame.

2. Add bldgs.

DATA (Current Workspace)   \student\1_modules1_5\baltimore_data
RESULTS (Scratch Workspace)   \student\1_modules1_5\baltimore_results

97

3. Search for and open the Intersect tool. Use it to intersect the buildings in Baltimore (bldgs) with each storm surge polygon (NED5flood). Name the output feature class **bldgsNED5**.

4. Repeat the process for LIDAR5flood. Name the file **bldgsLIDAR5**.

5. Open the attribute table for bldgsNED5. Record the number of buildings affected.

6. Repeat steps 3–5 for LIDAR5flood.

Q6 ***Which dataset has more buildings affected by the 5-foot surge? How many more houses are affected?***

7. Activate the Storm Surge 8 ft and Below data frame. Add bldgs. Repeat steps 2–5.

Q7 ***How many more houses are affected by the 8-foot storm surge when using the NED calculation?***

8. Activate the Storm Surge 12 ft and Below data frame. Add bldgs. Repeat steps 2–5.

Q8 ***How many more houses are affected by the 12-foot storm surge when using the NED calculation?***

9. Activate the Storm Surge 18 ft and Below data frame. Add bldgs. Repeat steps 2–5.

Q9 ***How many more houses are affected by the 18-foot storm surge when using the NED calculation?***

Q10 ***Why is there a difference in the number of houses affected when calculated using the two different datasets? Why might that difference increase with larger storm surges?***

Deliverable 1: A map comparing traditional 30-meter elevation data to lidar that also includes the number of buildings affected by the various storm surge delineations.

Draw conclusions and present the results

After creating your comparison map for flooding in Baltimore, choose a method of presenting your conclusions. Always keep the audience in mind as you prepare to report your results; they may not share your GIS expertise. Per FEMA's request, your results should be presented on ArcGIS Online.

MAKING
SPATIAL
DECISIONS
USING GIS
AND LIDAR

2

SURGING
SEAS

PROJECT 2

San Francisco, California

You can download a 30-meter DEM from the US Geological Survey National Map Viewer website: http://viewer.nationalmap.gov/viewer/.

Recommended deliverables

- **Deliverable 1:** A map comparing traditional 30-meter elevation data to lidar that also includes the number of buildings affected by the various storm surge delineations.

The questions asked in this project are both quantitative and qualitative. They identify key points that should be addressed in your analysis and final presentation.

Document your work, set environments, and examine the data

1. Open sanfran_basics1.mxd. Name the file **sanfran_elevation1.mxd**, and save it to the sanfrancisco_results folder.

2. Edit the map document properties as needed for this project.

This map document shows the construction of new FIRMs for FEMA.

DATA (Current Workspace)   \student\1_modules1_5\sanfrancisco_data
RESULTS (Scratch Workspace)   \student\1_modules1_5\sanfrancisco_results

99

MAKING
SPATIAL
DECISIONS
USING GIS
AND LIDAR

2

SURGING
SEAS

Analyze

Convert data to the correct projection and linear unit

1. Remove all layers.

2. Add SF_DEM.

SF_DEM is the raster that you made from the LAS dataset. You know that the map projection is NAD_1983_UTM_Zone_10N. You also know that the linear unit is in meters.

3. Zoom to a pixel, and measure its spatial dimension. Alternatively, go to the raster's properties and, on the Source dialog box, view the Cell Size (X,Y).

Q1   ***What is the spatial dimension of one pixel of SF_DEM?***

4. Add SF_elev_NED. Right-click SF_elev_NED, and go to Data > View Item Description.

Q2   ***What is the map projection and the linear unit?***

Q3   ***What is the spatial dimension of one pixel of SF_elev_NED?***

5. In the table of contents, select SF_DEM and rename it **LIDAR and SF_elev_NED to NED**.

Q4   ***How is the elevation range different in the two files?***

Use the Raster Calculator tool to establish surge areas

You are now ready to start making the FIRM maps. FEMA has established the following hurricane inundation zones. Use these values to make your FIRM maps (the storm surge heights have been converted to meters):

- Category I Hurricane   Storm Surge 0–5 ft   Storm Surge 0–1.5 m
- Category II Hurricane   Storm Surge 8 ft   Storm Surge 2.4 m
- Category III Hurricane   Storm Surge 12 ft   Storm Surge 3.7 m
- Category IV Hurricane   Storm Surge 18 ft   Storm Surge 5.5 m

1. Using the Raster Calculator tool, create storm surge rasters for both NED and LIDAR.

Remember: Create each storm surge in an individual data frame.

MAKING
SPATIAL
DECISIONS
USING GIS
AND LIDAR

2

SURGING
SEAS

2. Use the following Map Algebra expressions:

("NED" >= 0) & ("NED" <= 1.5)
("NED" >= 0) & ("NED" <= 2.4)
("NED" >= 0) & ("NED" <= 3.7)
("NED" >= 0) & ("NED" <= 5.5)

3. Name all output rasters appropriately, and save them to sanfrancisco_results\sanfrancisco.gdb.

Convert the flood surge rasters into polygons

1. First, for each flood surge raster, select Value = 1 in their respective raster attribute tables.

2. Use the Raster to Polygon tool to convert each individual raster to a polygon. In the conversion, do not select the Simplify Polygons option.

3. Name all output feature classes appropriately, and save them to the sanfrancisco_results \sanfrancisco.gdb\layers feature dataset.

Intersect the buildings with the storm surge polygons

1. Use the Intersect tool to intersect the buildings in San Francisco with each storm surge polygon. Name all output feature classes appropriately, and save them to the sanfrancisco_results\ sanfrancisco.gdb\layers feature dataset.

2. Open each attribute table, and answer the following questions.

Q5 *How many more buildings are affected by the 1.5-meter storm surge when using the NED calculation?*

Q6 *How many more buildings are affected by the 2.4-meter storm surge when using the lidar calculation?*

Q7 *How many more buildings are affected by the 3.7-meter storm surge when using the lidar calculation?*

DATA (Current Workspace)   \student\1_modules1_5\sanfrancisco_data
RESULTS (Scratch Workspace)   \student\1_modules1_5\sanfrancisco_results

101

MAKING
SPATIAL
DECISIONS
USING GIS
AND LIDAR

SURGING
SEAS

Q8 *How many more buildings are affected by the 5.5-meter storm surge when using the lidar calculation?*

Q9 *Why is there a difference in the number of buildings affected when calculated using the two different datasets?*

Another way to find the buildings in the storm surge polygons is to use the Select By Location tool. Set the target layer to bldgs, the source layer to the flood layers, and the spatial selection method to "intersect the source layer feature." When the appropriate buildings are selected, you can then export the data, name it appropriately, and save it to the designated folder.

Deliverable 1: A map comparing traditional 30-meter elevation data to lidar that also includes the number of buildings affected by the various storm surge delineations.

Draw conclusions and present the results

After creating your comparison map for flooding in San Francisco, choose a method of presenting your conclusions. Always keep the audience in mind as you prepare to report your results; they may not share your GIS expertise. Per FEMA's request, your results should be presented on ArcGIS Online.

MAKING
SPATIAL
DECISIONS
USING GIS
AND LIDAR

3

SURGING
SEAS

PROJECT 3

On your own

Scenario

You have worked through a guided project and repeated the analysis for another project. In this project, you will reinforce your skills by researching and analyzing a similar scenario entirely on your own. First, identify your study area and acquire data for your analysis. You may want to study a local area.

Here is a list of topics that have been studied using lidar. The following represent possible project ideas:

- Forest characterization—canopy height and density
- Flood modeling
- Finding faults
- Geomorphic mapping
- Stream slope
- Archaeology field campaigns
- Mining—calculation of ore volumes
- Wind farm optimization

MAKING
SPATIAL
DECISIONS
USING GIS
AND LIDAR

SURGING
SEAS

Many different websites distribute lidar data. Most of the available lidar data is within the United States. There are national data sites, and there are state data distributors. Reading the metadata and identifying a specific area of study is critical before downloading files. Some lidar files come compressed and require third-party software to convert them into the LAS format used by ArcMap.

- Open Topography facilitates community access to high-resolution, Earth science-oriented, topography data. It is a National Science Foundation-funded data facility. http://www.opentopography.org.
- The Earth Explorer provides online access to remotely sensed data from the US Geological Survey Earth Resources Observation and Science (EROS) Center archive. http://earthexplorer.usgs.gov/.
- National Oceanic and Atmospheric Administration, a world leader in coast science and management, provides state lidar datasets. http://www.csc.noaa.gov/dataviewer/#.
- Wikipedia lists national lidar datasets, organized according to state. http://en.wikipedia.org/wiki/National_LIDAR_Dataset_%E2%80%93_USA.

Research

Research the problem and answer the following questions:

1. What is the area of study?
2. What problem are you going to study?
3. What data is available?

MODULES 6 AND 7 SCENARIO

Modules 6 and 7 involve two geographic locations and a series of campus-based problems to be solved by an alliance between James Madison University and the University of San Francisco. The geographic science departments at each institution are interested in finding the optimal locations for solar panels on their rooftops. Researchers would like to produce roof-top heat maps that show the solar radiation incident on each rooftop. Before the universities can create the solar application, they must first manually fix errors in class codes and set class codes from features in their lidar data.

James Madison University obtained its lidar data from the William and Mary Center for Geospatial Analysis: https://www.wm.edu/as/cga/VALIDAR/index.php.

The building footprints were obtained from VGIN 2011. VGIN is the Virginia Geographic Information Network: http://www.vita.virginia.gov/.

The University of San Francisco obtained its lidar data from the US Geological Survey Earth Explorer website: http://earthexplorer.usgs.gov/.

Building footprints were obtained from the SF OpenData website: https://data.sfgov.org/Other/Data-Catalog/h4ui-ubbu.

The university alliance divided the lidar project into two parts:

- **Module 6: Corrected 3D campus modeling**
- **Module 7: Location of solar panels**

Note: Modules 6 and 7 use the same data; therefore, there is one folder for jmu_data and one folder for jmu_results in the exercise data. Similarly, there is one folder for sfu_data and sfu_results.

CORRECTED 3D CAMPUS MODELING

Projects included in this module

- **Project 1:** James Madison University, Harrisonburg, Virginia
- **Project 2:** University of San Francisco, San Francisco, California
- **Project 3:** On your own

Student worksheets

Project 1: James Madison University, Harrisonburg, Virginia

- File name: 6a_JMU_campus_worksheet.docx
- Location: EsriPress\MSDLidar\student\2_modules6_7\1class_codes\project1_documents

Project 2: University of San Francisco, San Francisco, California

- File name: 6b_SFU_campus_worksheet.docx
- Location: EsriPress\MSDLidar\student\2_modules6_7\1class_codes\project2_documents

Objectives

- Produce an accurate 3D dataset.
- Manually fix any errors in the class codes.
- Set the class codes from features.
- Visualize the campus in 3D using ArcScene.
  - Visualize the campus as a surface.
  - Visualize the campus as a contour.

MAKING
SPATIAL
DECISIONS
USING GIS
AND LIDAR

CORRECTED
3D CAMPUS
MODELING

James Madison University, Harrisonburg, Virginia

Recommended deliverables

- **Deliverable 1:** A map of the study area showing campus lidar with class codes corrected using features.

The questions asked in this project are both quantitative and qualitative. They identify key points that should be addressed in your analysis and final presentation.

Document your work, set environments, and examine the data

1. Open ArcMap. Edit the map document properties as needed for this project. Store relative pathnames.

For this module and the next, **jmu_data** will be your project folder (EsriPress\MSDLidar\ student\2_modules6_7\jmu_data). Make sure that it is stored in a place where you have write access. You can store your project results in the **jmu_results** folder (EsriPress\MSDLidar\ student\2_modules6_7\jmu_results).

2. Set environments:
   - Open the data frame properties. Set the map projection to Projected Coordinate Systems > State Plane > NAD_1983 HARN (US Feet) > NAD_1983_HARN_StatePlane_Virginia_North_ FIPS_4501.
   - On the Geoprocessing menu, click Environments.
   - Expand Workspace, and set the current workspace to \student\2_modules6_7\jmu_data.

DATA (Current Workspace)   \student\2_modules6_7\jmu_data
RESULTS (Scratch Workspace) \student\2_modules6_7\jmu_results

- Set the scratch workspace to \student\2_modules6_7\jmu_results.
- For the output coordinate system, select Same as Display.

3. Name the map document **JMUsolar1**, and save it to \student\2_modules6_7\jmu_results.

MAKING
SPATIAL
DECISIONS
USING GIS
AND LIDAR

1

CORRECTED
3D CAMPUS
MODELING

Analyze

1. Add the study_area feature class and make it hollow. Add the Imagery with Labels basemap.

Q1 *Describe the landscape in the study area.*

The changes that you make when you reclassify the LAS dataset cannot be undone. They are permanent. **Do not use the LAS files in the following folder: \2_modules6_7\jmu_data\ ARCHIVE_LAS_files.** Use the LAS dataset in \student\2_modules6_7\jmu_results to complete this exercise.

2. Add the reclassify.lasd dataset from the jmu_results folder.

3. Turn on the LAS Dataset toolbar. Change the point display option from Elevation to Class.

Remember: You are not seeing all of the lidar points. The data percentage shown in the table of contents gives the fraction of the points displayed. You can increase the number of points displayed by performing the following step.

4. Right-click the reclassify.lasd dataset and go to Properties:
   - Click the Display tab.
   - Increase the point limit to **5000000**.
   - Select the "Use scale to control full resolution" check box. Change the Full Resolution Scale to **5000**.
   - Move the Point Density slider to Fine.
   - (Optional) Select the "Always display the LAS file extents" and "Display LAS file names" check boxes.

Every lidar point in this dataset has a defining classification assigned to it. In this dataset, you can see that some points are not classified properly. ArcMap has tools to change class codes. You will reclassify the LAS dataset using two geoprocessing tools. The first tool is Set LAS Class Codes Using Features. This tool changes lidar classification codes assigned to lidar points based on their proximity to a feature dataset. The second tool is Change LAS Class Codes. This tool will reclassify one set of classification codes into another.

MAKING
SPATIAL
DECISIONS
USING GIS
AND LIDAR

1

CORRECTED
3D CAMPUS
MODELING

The following table shows the ASPRS (American Society for Photogrammetry and Remote Sensing) lidar point classes.

ASPRS STANDARD LIDAR POINT CLASSES

Classification Value (Bits 0–4)	Meaning
0	Never classified
1	Unassigned
2	Ground
3	Low Vegetation
4	Medium Vegetation
5	High Vegetation
6	Building
7	Noise
8	Model Key
9	Water
10	Reserved for ASPRS Definition
11	Reserved for ASPRS Definition
12	Overlap
13–31	Reserved for ASPRS Definition

5. Add the pond feature class. You can now see that the lidar points within the pond are not classified as water, which is 9.

6. Search for and open the Set LAS Class Codes Using Features tool. Enter the following settings:
   • Set the input LAS dataset to reclassify.lasd.
   • Set the input feature class to pond.
   • Set the buffer distance to **1**.
   • Set the new class to **9**.

DATA (Current Workspace)   \student\2_modules6_7\jmu_data
RESULTS (Scratch Workspace)   \student\2_modules6_7\jmu_results

The reclassification changes the codes in the LAS dataset. All modifications to the classification codes in LAS files are permanent.

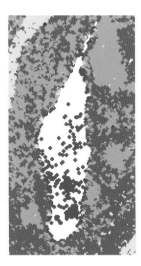

MAKING
SPATIAL
DECISIONS
USING GIS
AND LIDAR

1

CORRECTED
3D CAMPUS
MODELING

The gray points appear to represent buildings.

7. Using the Identify tool, click the points that might represent buildings.

8. Add the bldgs feature dataset and repeat step 6, but set the new class to **6**.

9. Remove the pond and bldgs feature classes.

10. Turn on the Imagery with Labels basemap and zoom to where you now see gray areas. These areas seem to represent vegetation.

Q2 **What number represents the gray color?**

11. Search for and open the Change LAS Class Codes tool. Enter the following settings:
    - Set the input LAS dataset to reclassify.lasd.
    - Set the class codes to **1**. (Press the + button.)
    - Set the new class to **4** (for medium vegetation).

Q3 **What are the two areas in the lower southeast corner that have no lidar coverage?**

Q4 **What is the class code of the dark gray strips?**

MAKING
SPATIAL
DECISIONS
USING GIS
AND LIDAR

CORRECTED
3D CAMPUS
MODELING

12. Repeat step 11, but change the class code 11 to **2** for ground.

13. Go to Catalog > Properties > Statistics > Calculate and click reclassify.lasd.

Q5 **What are the classification codes?**

14. When asked if you want to save the changes, click Yes.

15. Construct a map layout using good cartographic principles.

16. Save the map document as **JMUsolar1.mxd**. Save the map document again as **JMUsolar2.mxd**.

Deliverable 1: A map of the study area showing campus lidar with class codes corrected using features.

Draw conclusions and present the results

Once you have created your corrected campus map for James Madison University, choose a method of presenting your conclusions. Always keep the audience in mind as you prepare to report your results; they may not share your GIS expertise. Your results can be presented in a Word document, a PowerPoint presentation, or a more technical presentation mode, such as ArcGIS Online.

MAKING
SPATIAL
DECISIONS
USING GIS
AND LIDAR

2

CORRECTED
3D CAMPUS
MODELING

PROJECT 2

University of San Francisco, San Francisco, California

Recommended deliverables

- **Deliverable 1:** A map of the study area showing campus lidar with class codes corrected using features.

The questions asked in this project are both quantitative and qualitative. They identify key points that should be addressed in your analysis and final presentation.

Document your work, set environments, and examine the data

1. Open a new ArcMap document. Edit the map document properties as needed for this project. Store relative pathnames.

For this module and the next, **sfu_data** will be your project folder (EsriPress\MSDLidar \student\2_modules6_7\sfu_data). Make sure that it is stored in a place where you have write access. You can store your project results in the **sfu_results** folder (EsriPress\MSDLidar \student\2_modules6_7\sfu_results).

2. Set environments:
   - Open the data frame properties. Set the map projection to Projected Coordinate Systems > NAD_1983_UTM_Zone_10N.
   - On the Geoprocessing menu, click Environments.
   - Expand Workspace, and set the current workspace to \student\2_modules6_7\sfu_data.
   - Set the scratch workspace to \student\2_modules6_7\sfu_results.
   - For the output coordinate system, select Same as Display.

3. Name the map document **SFUsolar1**, and save it to \2_module6_7\sfu_results.

DATA (Current Workspace)   \student\2_modules6_7\sfu_data
RESULTS (Scratch Workspace)   \student\2_modules6_7\sfu_results

MAKING
SPATIAL
DECISIONS
USING GIS
AND LIDAR

2

CORRECTED
3D CAMPUS
MODELING

Analyze

1. Add the sa_sanfrancisco feature class and make it hollow. Add the Imagery with Labels basemap.

Q1   Describe the landscape in the study area.

The changes that you make when you reclassify the LAS dataset cannot be undone. They are permanent. **Do not use the LAS files in the following folder: \2_modules6_7\sfu_data\ ARCHIVE_LAS_files**. Use the LAS dataset in \student\2_modules6_7\sfu_results\LAS_files to complete this exercise.

2. Add the reclassify.lasd dataset from the sfu_results folder.

3. Turn on the LAS Dataset toolbar. Change the point display option from Elevation to Class.

4. Set the display to the maximum size:
    *   Set the point limit to **2800000**.
    *   Set the full resolution scale to **5000**.
    *   Move the Point Density slider to Fine.

5. Using the Set LAS Class Codes Using Features tool, add bldgs, use a buffer distance of **1**, and use a new class of **6**.

6. Turn on the basemap and zoom to where you see green areas. The green areas seem to represent vegetation.

Q2   What class code represents the brown color?

7. Use the Change LAS Class Codes tool to change the gray color of 1 (Unassigned) to **2** for ground.

8. Add bldgs. Open the Set LAS Class Codes Using Features tool. Input the bldgs feature class with a buffer distance of **1**. Set the new class to **6**.

DATA (Current Workspace)   \student\2_modules6_7\sfu_data
RESULTS (Scratch Workspace)   \student\2_modules6_7\sfu_results

MAKING
SPATIAL
DECISIONS
USING GIS
AND LIDAR

2

CORRECTED
3D CAMPUS
MODELING

Q3　**What are the classification codes?**

9. Construct a map layout using good cartographic principles.

10. Save the map document as **SFUsolar1.mxd**. Save the map document again as **SFUsolar2.mxd**.

Deliverable 1: A map of the study area showing campus lidar with class codes corrected using features.

Draw conclusions and present the results

Once you have created your corrected campus map for the University of San Francisco, choose a method of presenting your conclusions. Always keep the audience in mind as you prepare to report your results; they may not share your GIS expertise. Your results can be presented in a Word document, a PowerPoint presentation, or a more technical presentation mode, such as ArcGIS Online.

MAKING
SPATIAL
DECISIONS
USING GIS
AND LIDAR

CORRECTED
3D CAMPUS
MODELING

PROJECT **3**

On your own

Scenario

You have worked through a guided project and repeated the analysis for another project. In this project, you will reinforce your skills by researching and analyzing a similar scenario entirely on your own. First, identify your study area and acquire data for your analysis. You may want to study a local area.

Here is a list of topics that have been studied using lidar. The following represent possible project ideas:

- Forest characterization—canopy height and density
- Flood modeling
- Finding faults
- Geomorphic mapping
- Stream slope
- Archaeology field campaigns
- Mining—calculation of ore volumes
- Wind farm optimization

Many different websites distribute lidar data. Most of the available lidar data is within the United States. There are national data sites, and there are state data distributors. Reading the metadata and identifying a specific area of study is critical before downloading files. Some lidar files come compressed and require third-party software to convert them into the LAS format used by Esri.

- Open Topography facilitates community access to high-resolution, Earth science-oriented, topography data. It is a National Science Foundation-funded data facility. http://www.opentopography.org.
- The Earth Explorer provides online access to remotely sensed data from the US Geological Survey Earth Resources Observation and Science (EROS) Center archive. http://earthexplorer.usgs.gov/.
- National Oceanic and Atmospheric Administration, a world leader in coast science and management, provides state lidar datasets. http://www.csc.noaa.gov/dataviewer/#.
- Wikipedia lists national lidar datasets, organized according to state. http://en.wikipedia.org/wiki/National_LIDAR_Dataset_%E2%80%93_USA.

Research

Research the problem and answer the following questions:

1. What is the area of study?
2. What problem are you going to study?
3. What data is available?

MAKING
SPATIAL
DECISIONS
USING GIS
AND LIDAR

CORRECTED
3D CAMPUS
MODELING

MODULE 7

LOCATION OF SOLAR PANELS

Projects included in this module

- **Project 1:** James Madison University, Harrisonburg, Virginia
- **Project 2:** University of San Francisco, San Francisco, California
- **Project 3:** On your own

Student worksheets

Project 1: James Madison University, Harrisonburg, Virginia
- File name: 7a_JMU_solar_worksheet.docx
- Location: EsriPress\MSDLidar\student\2_modules6_7\2solar\project1_documents

Project 2: University of San Francisco, San Francisco, California
- File name: 7b_SFU_solar_worksheet.docx
- Location: EsriPress\MSDLidar\student\2_modules6_7\2solar\project2_documents

Objectives

- Produce rooftop heat maps.
- Convert the LAS dataset to raster.
- Run the Extract by Mask tool using building footprints.
- Run the Solar Radiation tool to calculate monthly solar radiation received by extracted lidar rooftops.
- Produce an optimal solar radiation map showing roof classification for solar panel installation.

MAKING
SPATIAL
DECISIONS
USING GIS
AND LIDAR

1

LOCATION OF
SOLAR PANELS

PROJECT 1

James Madison University, Harrisonburg, Virginia

Recommended deliverables

- **Deliverable 1:** A map of the study area showing campus buildings with extracted raster lidar.
- **Deliverable 2:** A map of the study area showing solar radiation calculated using lidar for the campus. Include insets for specific buildings.

The questions asked in this project are both quantitative and qualitative. They identify key points that should be addressed in your analysis and final presentation.

Document your work, set environments, and examine the data

1. Open ArcMap. Edit the map document properties as needed for this project. Store relative pathnames.

The folder structure and environments for module 7 are the same as those in module 6.

Analyze starting with a basemap

Extract building lidar

1. Open JMUsolar2.mxd and remove the Imagery with Labels basemap and Reference layer.

2. Search for and open the LAS Dataset to Raster tool, which converts the LAS dataset to a raster dataset. Enter the following settings:
   - Set the input LAS dataset to reclassify.
   - Set the output raster to \jmu_results\jmu_results.gdb**allpts_raster**.

MAKING
SPATIAL
DECISIONS
USING GIS
AND LIDAR

1

LOCATION OF
SOLAR PANELS

- Set the value field to ELEVATION.
- For the interpolation type, choose Binning. Set the cell assignment type to MAXIMUM, and set the void fill method to NATURAL_NEIGHBOR.
- Set the output data type to INT.
- Accept the remaining default optional values.
- Click OK.

3. Remove the reclassify.lasd and study_area data layers.

Incoming solar radiation (insolation) is the primary energy source that drives many of the earth's physical and biological processes. Topography is a major factor that determines the spatial variability of solar insolation. Variation in elevation, orientation (slope and aspect), and shadows cast by topographic features all affect the amount of insolation received at different locations. This variability also changes with time of day and year.

Lidar is uniquely suited for accurate analysis of solar radiation (insolation) because it contains high accuracy and precision elevation data, which can be used to derive slope and aspect, among other datasets. You will now extract the lidar returns for each building rooftop and then run the Solar Radiation tool to determine the optimal location for solar panel installation.

4. Search for and open the Extract by Mask tool. Enter the following settings:
   - Set the input raster to allpts_raster.
   - Set the input raster or feature mask data to \jmu_results\jmu.gdb\Layers\bldgs.
   - Set the output raster to \jmu_results\jmu_results.gdb**bldgs_raster**.

5. Remove allpts_raster.

Q1   *What are the lowest and highest building elevations?*

6. Save the map document.

7. Create a map layout using good cartographic principles.

8. Save the map document again as **JMUsolar3.mxd**.

Deliverable 1: A map of the study area showing campus buildings with extracted raster lidar.

Calculate the amount of solar radiation received on the James Madison University campus

The tools in the Solar Radiation toolset in the ArcGIS Spatial Analyst extension allow you to map and analyze the effects of the sun over a geographic area for specific time periods. The tools account for atmospheric effects, site latitude and elevation, steepness (slope) and compass direction (aspect), daily and seasonal shifts of the sun angle, and effects of shadows cast by surrounding topography.

1. Open JMUsolar3.mxd and display it in data view. Remove the basemap.

Using area solar radiation analysis, the insolation is calculated for the entire study area in Wh/m^2 (watt-hours per square meter). The highest amounts of insolation are shown in red, and the lowest are shown in blue.

You will calculate the amount of solar radiation that each building rooftop will receive on December 21, the winter solstice in the Northern Hemisphere, when sun angles are at their minimum. Feel free to experiment with other dates.

2. Search for and open the Area Solar Radiation tool. Enter the following settings:
    *   Set the input raster to bldgs_raster.
    *   Set the output global radiation raster to \jmu_results\jmu.gdb**winter.**
    *   Set the time configuration to within a day. Pick December 21, which is day 355.
    *   Accept all other optional default values.
    *   Click OK.

Q2   ***How much solar energy is incident in the red areas (high) in Wh/m^2 on December 21? How much in the blue areas (low)?***

Q3   ***In which areas would it be best to install solar panels? Why? What other considerations affect where you can place solar panels?***

3. Add the Imagery basemap and zoom to the lower left (southeast) corner of the scene.

4. Change the name of the data frame from Layers to **Campus**.

5. Remove bldgs_raster.

DATA (Current Workspace)   \student\2_modules6_7\jmu_data
RESULTS (Scratch Workspace)   \student\2_modules6_7\jmu_results

Calculate the amount of solar radiation for dormitories

1. Copy the Campus data frame and paste it into the table of contents. Change the name of the pasted data frame to **Dormitory**.

2. Activate the Dormitory data frame.

3. Add the dormitory feature class from \jmu_data\JMU.gdb\Layers.

4. Search for and open the Extract by Mask tool. Enter the following settings:
   • Set the input raster to winter.
   • Set the input raster or feature mask data to dormitory.
   • Set the output raster to \jmu_results\JMU_results.gdb**dorm**.

5. Remove the winter raster. Zoom to the dorm raster layer.

6. Make dormitory hollow with an outline width of 3.

7. Right-click dorm, and go to Properties. On the Symbology tab, under Color Ramp, select an appropriate color ramp (for example, blue-to-red ramp).

8. Turn on the Draw toolbar and draw a circle around the area on each dorm that would be the best place for a solar panel. Make the circles hollow with an outline of 3.

Q4 **How many Wh/m² of solar energy are incident on December 21 in the red areas? How many in the blue areas?**

9. Create a layout of the campus solar radiation with an inset of the dormitories data frame using appropriate cartographic principles.

10. Save the JMUsolar3.mxd map.

Q5 **Based on your analysis, which buildings would be the best sites for solar panels? Which would be the worst sites? Why?**

Q6 **When you zoom to the building areas that seem most suitable for the solar panels, what features do you observe?**

Deliverable 2: A map of the study area showing solar radiation calculated using lidar for the campus. Include insets for specific buildings.

Draw conclusions and present the results

Once you have created the radiation maps for James Madison University, choose a method of presenting your conclusions. Always keep the audience in mind as you prepare to report your results; they may not share your GIS expertise. Your results can be presented in a Word document, a PowerPoint presentation, or a more technical presentation mode, such as ArcGIS Online.

DATA (Current Workspace)   \student\2_modules6_7\jmu_data
RESULTS (Scratch Workspace)   \student\2_modules6_7\jmu_results

MAKING
SPATIAL
DECISIONS
USING GIS
AND LIDAR

2

LOCATION OF
SOLAR PANELS

PROJECT 2

San Francisco University, San Francisco, California

Recommended deliverables

- **Deliverable 1:** A map of the study area showing campus buildings with extracted raster lidar.
- **Deliverable 2:** A map of the study area showing solar radiation calculated using lidar for the campus. Include insets for specific buildings.

The questions asked in this project are both quantitative and qualitative. They identify key points that should be addressed in your analysis and final presentation.

Document your work, set environments, and examine the data

1. Open ArcMap. Edit the map document properties as needed for this project. Store relative pathnames.

The folder structure and environments for module 7 are the same as those in module 6.

Analyze starting with a basemap

Extract building lidar

1. In ArcMap, open SFUsolar2.mxd. Remove Imagery with Labels.

2. Search for and open the LAS Dataset to Raster tool. Enter the following settings:
    - Set the input LAS dataset to \sfu_results\reclassify.
    - Set the output raster to \sfu_results\sfu_results.gdb**allpts_raster**.
    - Set the value field to ELEVATION.

DATA (Current Workspace)   \student\2_modules6_7\sfu_data
RESULTS (Scratch Workspace)   \student\2_modules6_7\sfu_results

MAKING
SPATIAL
DECISIONS
USING GIS
AND LIDAR

2

LOCATION OF
SOLAR PANELS

- For the interpolation type, choose Binning. Set the cell assignment type to MAXIMUM, and set the void fill method to NATURAL_NEIGHBOR.
- Set the output data type to INT.
- Accept the remaining default optional values.
- Click OK.

3. Remove the reclassify.lasd and sa_sanfrancisco study_area data layers.

Incoming solar radiation (insolation) is the primary energy source that drives many of the earth's physical and biological processes. Topography is a major factor that determines the spatial variability of insolation. Variation in elevation, orientation (slope and aspect), and shadows cast by topographic features all affect the amount of insolation received at different locations. This variability also changes with time of day and year.

Lidar is uniquely suited for accurate analysis of solar radiation (insolation) because it contains high accuracy and precision elevation data, which can be used to derive slope and aspect, among other datasets. You will now extract the lidar returns for each building rooftop and then run the Solar Radiation tool to determine the optimal location for solar module installation.

4. Search for and open the Extract by Mask tool. Enter the following settings:
   - Set the input raster to allpts_raster.
   - Set the input raster or feature mask data to bldgs.
   - Set the output raster to \sfu_results\SFU_results.gdb**bldgs_raster**.

5. Remove allpts_raster.

Q1   *What are the lowest and highest building elevations?*

6. Save the map document.

7. Create a map layout using good cartographic principles.

8. Save the map document again as **SFUsolar3.mxd**.

Deliverable 1: A map of the study area showing campus buildings with extracted raster lidar.

Calculate the amount of solar radiation received on the University of San Francisco campus

The tools in the Solar Radiation toolset in the ArcGIS Spatial Analyst extension allow you to map and analyze the effects of the sun over a geographic area for specific time periods. The

DATA (Current Workspace)   \student\2_modules6_7\sfu_data
RESULTS (Scratch Workspace)   \student\2_modules6_7\sfu_results

MAKING
SPATIAL
DECISIONS
USING GIS
AND LIDAR

2

LOCATION OF
SOLAR PANELS

tools account for atmospheric effects, site latitude and elevation, steepness (slope) and compass direction (aspect), daily and seasonal shifts of the sun angle, and effects of shadows cast by surrounding topography.

1.   Open SFUsolar3.mxd and remove the basemap.

Using area solar radiation analysis, the insolation is calculated for the entire study area in Wh/m^2 (watt-hours per square meter). The highest amounts of insolation are shown in red, and the lowest are shown in blue.

You will calculate the amount of solar radiation that each building rooftop will receive on December 21, the winter solstice in the Northern Hemisphere, when sun angles are at their minimum. Explore other dates to see the annual variation in insolation.

2.   Search for and open the Area Solar Radiation tool. Enter the following settings:
   - Set the input raster to bldgs_raster.
   - Set the output global radiation raster to \sfu_results\SFUresults.gdb**winter.**
   - Set the time configuration to within a day. Pick December 21, which is day 355.
   - Accept all other optional default values.
   - Click OK.

Q2   ***How much solar energy is incident in the red areas in Wh/m^2 on December 21? How much in the blue areas?***

Q3   ***In which areas would it be best to install solar panels? Why? What other considerations affect where you can install solar panels on a campus building?***

3.   Add the Imagery basemap and zoom to the lower left (southeast) corner of the scene.

4.   Change the name of the data frame from Layers to **Campus**.

5.   Remove bldgs_raster.

Calculate the amount of solar radiation for dormitories

1.   Copy the Campus data frame and paste it into the table of contents. Change the name of the pasted data frame to **Dormitory**.

2.   Activate the Dormitory data frame.

MAKING
SPATIAL
DECISIONS
USING GIS
AND LIDAR

2

LOCATION OF
SOLAR PANELS

3. Add the main_univ feature class.

4. Search for and open the Extract by Mask tool. Enter the following settings:
   - Set the input raster to winter.
   - Set the input raster or feature mask data to main_univ.
   - Set the output raster to \sfu_results\sfu_results.gdb**main_univ**.

5. Remove the winter raster. Zoom to the main_univ_dorm layer.

6. Make main_univ hollow with an outline width of 3.

7. Right-click dorm, and go to Properties. On the Symbology tab, under Color Ramp, select an appropriate color ramp (for example, the blue-to-red ramp).

8. Turn on the Draw toolbar and draw a circle around the area on the main university dorm building that would be the best location for a solar panel. Make the circle hollow with an outline of 3.

Q4   ***How much solar energy is incident in the red areas in Wh/m² on December 21? How much in the blue areas?***

9. Create a layout of the University of San Francisco campus solar radiation. Add an inset of the University of San Francisco dorm. Use appropriate cartographic principles.

10. Save the map as **SFUsolar3.mxd**.

Deliverable 2: A map of the study area showing solar radiation calculated using lidar for the campus. Include insets for specific buildings.

Draw conclusions and present the results

Once you have created the radiation maps for the University of San Francisco, choose a method of presenting your conclusions. Always keep the audience in mind as you prepare to report your results; they may not share your GIS expertise. Your results can be presented in a Word document, a PowerPoint presentation, or a more technical presentation mode, such as ArcGIS Online.

MAKING
SPATIAL
DECISIONS
USING GIS
AND LIDAR

3

LOCATION OF
SOLAR PANELS

PROJECT 3

On your own

Scenario

You have worked through a guided project and repeated the analysis for another project. In this project, you will reinforce your skills by researching and analyzing a similar scenario entirely on your own. First, identify your study area and acquire data for your analysis. You may want to study a local area.

Here is a list of topics that have been studied using lidar. The following represent possible project ideas:

- Forest characterization—canopy height and density
- Flood modeling
- Finding faults
- Geomorphic mapping
- Stream slope
- Archaeology field campaigns
- Mining—calculation of ore volumes
- Wind farm optimization

Many different websites distribute lidar data. Most of the available lidar data are within the United States. There are national data sites, and there are state data distributors. Reading the metadata and identifying a specific area of study is critical before downloading files. Some lidar files come compressed and require third-party software to convert them into the LAS format used by Esri.

- Open Topography facilitates community access to high-resolution, Earth science-oriented, topography data. It is a National Science Foundation-funded data facility. http://www.opentopography.org.

- The Earth Explorer provides online access to remotely sensed data from the US Geological Survey Earth Resources Observation and Science (EROS) Center archive. http://earthexplorer.usgs.gov/.

- National Oceanic and Atmospheric Administration, a world leader in coast science and management, provides state lidar datasets. http://www.csc.noaa.gov/dataviewer/#.

- Wikipedia lists national lidar datasets, organized according to state. http://en.wikipedia.org/wiki/National_LIDAR_Dataset_%E2%80%93_USA.

Research

Research the problem and answer the following questions:

1. What is the area of study?
2. What problem are you going to study?
3. What data is available?

SHORELINE CHANGE AFTER HURRICANE SANDY

SCENARIO ···

The US Geological Survey (USGS) has contracted your GIS Company to analyze the change of the New Jersey and New York coasts after Hurricane Sandy. Pre- and post-Sandy lidar data has been obtained from the Digital Coast, a website sponsored by the National Oceanic and Atmospheric Administration (NOAA). These two sets of lidar data were used to compare the most common types of coastal impact, which are the following:

- Erosion
- Deposition
- Island breaching

USGS wants pre- and post-event elevation maps that show elevation decline from erosion, shoreline retreat, and elevation deposition for wave and surge motion that has changed the shoreline.

The inspiration for this module came from the USGS 3D lidar topography of New Jersey acquired before and after Hurricane Sandy: http://coastal.er.usgs.gov/hurricanes/sandy/lidar/newjersey.php.

Projects included in this module

- **Project 1:** Deal, New Jersey
- **Project 2:** Pelham Manor, New York
- **Project 3:** On your own

Student worksheets

Project 1: Deal, New Jersey

- File name: 8a_NJ_Sandy_worksheet.docx
- Location: EsriPress\MSDLidar\student\3_module8\nj_documents

Project 2: Pelham Manor, New York

- File name: 8b_NY_Sandy_worksheet.docx
- Location: EsriPress\MSDLidar\student\3_module8\ny_documents

Objectives

- Prepare the LAS datasets for the pre- and post-Sandy New Jersey coastline.
- Analyze the raster datasets for the pre- and post-Sandy New Jersey coastline.
- Analyze the temporal change between the pre- and post-Sandy datasets.

MAKING
SPATIAL
DECISIONS
USING GIS
AND LIDAR

1

SHORELINE
CHANGE AFTER
HURRICANE
SANDY

PROJECT 1

Deal, New Jersey

Recommended deliverables

- **Deliverable 1:** A basemap of Deal, New Jersey.
- **Deliverable 2:** A map containing two data frames showing pre- and post-storm lidar elevation.
- **Deliverable 3:** A map showing the difference in elevation between pre- and post-Sandy lidar data.

The questions asked in this project are both quantitative and qualitative. They identify key points that should be addressed in your analysis and final presentation.

Document your work, set environments, and examine the data

1. Open ArcMap. Add descriptive map document properties to your map document.

2. Set environments:
   - Open the data frame properties. Set the map projection to Geographic Coordinate Systems > North America > NAD 1983.
   - On the Geoprocessing menu, click Environments.
   - Expand Workspace, and set the current workspace to \student\3_module8\nj_data.
   - Set the scratch workspace to \student\3_module8\nj_results.
   - For the output coordinate system, select Same as Display.

3. Save the map document as **nj_1** to the nj_results folder.

DATA (Current Workspace)   \student\3_module8\nj_data
RESULTS (Scratch Workspace)   \student\3_module8\nj_results

MAKING
SPATIAL
DECISIONS
USING GIS
AND LIDAR

SHORELINE
CHANGE AFTER
HURRICANE
SANDY

▌ Analyze starting with a basemap

1. Add NJ, study_area, study_area2, and places. Make study_area and study_area2 hollow.

This coastline map will look better if it is rotated.

2. Right-click Layers, and go to Properties. On General tab, set the rotation to **–75**.

3. Add the Imagery basemap. Note: The Imagery with Labels basemap is not a good choice here because the reference labels will not rotate.

4. Create a locational map and save it as **nj_1**. Save the map document again as **nj_2**.

Q1 ***Describe the physical and demographic characteristics of Deal, New Jersey.***

Deliverable 1: A basemap of Deal, New Jersey.

Compare pre- and post-Sandy lidar data

1. Open nj_2.mxd and remove all layers except study_area, study_area2, and the basemap.

2. Change the name of the data frame to **PRE 2010**.

3. In Catalog, create a LAS dataset inside the project1_NJ_results folder. Name the new dataset **PRE**.

4. Go to the PRE.lasd properties. Enter the following settings:
    - On the LAS Files tab, add LAS files from nj_results\LAS_2010_PRE\20100828_2010_ncmp_nj_05_geoclassified_ld_p10.
    - Click Statistics and Calculate.
    - Set x,y coordinate system to Geographic Coordinate Systems > North America > NAD 1983.
    - Set the z coordinate system to Vertical Coordinate Systems > North America > NAVD 1988.

The metadata for the LAS file is in the LAS_2010_PRE folder. The metadata specifies that the elevation is given in meters.

5. Drag the PRE.lasd dataset to the table of contents.

DATA (Current Workspace)   \student\3_module8\nj_data
RESULTS (Scratch Workspace)   \student\3_module8\nj_results

6. Go to the PRE.lasd properties. Go to Display, and set the following parameters:
   - Set the point limit to **5000000**.
   - Select the "Use scale to control full resolution" check box.
   - Move the Point Density slider to Fine.
   - Set the full resolution scale to **5000**.

7. Save your map document, and then repeat steps 3–6 for the post-Sandy data. Name the LAS dataset **POST.lasd,** and then add the LAS file 20121120_2012_PostSandy_NJ_000278_GeoClassified.las.

8. Save your map document again. Turn on the Effects toolbar and use the Swipe tool to alternate between images. In the Effects toolbar menu, make sure that POST.lasd is the selected file.

9. On the LAS Dataset toolbar, set the filters to Ground in both the PRE and the POST LAS datasets.

Q2   ***Describe any visible changes. Pay particular attention to the area around the Deal Casino. How has the shoreline changed there? (**Hint: **Look up the location of the Deal Casino.)***

10. Click the arrow next to the Add Data button and add data from ArcGIS Online. Search for Hurricane Sandy USGS. Add the 2012 Hurricane Sandy USGS Imagery. Click noaa_01nov2012_utm18 .mos, which will give you access to imagery along the damaged New Jersey coastline.

11. Insert a new data frame and name it **POST 2012**. Add the POST.lasd dataset. Remove the POST .lasd from the PRE 2010 data frame's table of contents.

12. Create a map containing two data frames with the PRE and POST lidar elevations.

13. Save the map document as **nj_2** to the nj_results folder. Save the document again as **nj_3** to the nj_results folder.

Deliverable 2: A map containing two data frames showing pre- and post-storm lidar elevation.

MAKING
SPATIAL
DECISIONS
USING GIS
AND LIDAR

SHORELINE
CHANGE AFTER
HURRICANE
SANDY

Calculate the elevation difference between pre- and post-Sandy lidar data

1. Open a new ArcMap document and add document properties. Store relative pathnames.

2. Add the study_area2 feature class.

3. Set environments:
   - Open the data frame properties. Set the map projection to NAD 1983 UTM Zone 18N.
   - On the Geoprocessing menu, click Environments.
   - Set the current workspace to \student\3_modules8\nj_data.
   - Set the scratch workspace to \student\3_modules8\nj_results.
   - For the output coordinate system, select Same as Display.
   - Set the processing extent to Same as layer study_area2.
   - Set the raster analysis mask to study_area2.

4. Save the map document as **nj3.mxd** to the nj_results folder.

5. Search for and open the LAS to Multipoint tool.

This tool imports one or more files in LAS format into a new multipoint feature class. The tool allows the user to read the lidar data file and load it into a geodatabase for further analysis. The ground spacing is acquired from the metadata.

6. Enter the following settings into the LAS to Multipoint tool:
   - From the LAS_PRE_folder, add the 20100828_2010_ncmp_nj_05_geoclassified_ld_p10.las file.
   - Set the output feature class to nj_results\NJ_Sandy.gdb\UTM**PRE_ALL.**
   - Set the point spacing to **8.748e-006 or .000008.**
   - Accept the remaining default values.
   - Click OK.

7. On the Geoprocessing menu, click Clip to open the Clip tool. Enter the following settings:
   - Set the input features to PRE_ALL.
   - Set the clip features to study_area2.
   - Set the output feature class to \nj_results\NJ_Sandy.gdb\UTM**PRE_ALL_SA.**
   - Click OK.

8. Remove PRE_ALL.

9. Search for and open the IDW interpolation tool. IDW stands for inverse distance weighted.

This tool interpolates a raster surface from points using an inverse distance weighted technique. This tool produces the best results when sampling is sufficiently dense (as in lidar data) regarding local variation.

10. Enter the following settings:
   - Set the input point features to PRE_ALL_SA.
   - Set the Z-value field to Shape.Z.
   - Set the output raster to \nj_results**PRE_ALL.** (Note: The raster is stored in the results folder and not the geodatabase.)
   - Set the output cell size to **2**.
   - Accept the remaining default values.
   - Click OK.

11. Remove PRE_ALL_SA. Change the study_area2 symbology to hollow.

12. Right-click PRE_ALL_SA, and go to Properties. Click the Symbology tab.

13. Use Defined Interval as the classification method. Set the intervals to **5**, **10**, **15**, **20**, and **24**.

14. Right-click PRE_ALL_SA, and go to Save As Layer File. Save the layer as **elevation** in the nj_results folder.

15. Repeat steps 5–11 for the POST LAS files, using the following guidelines:
   - When executing LAS to Multipoint, add the following two LAS files from LAS_2012_POST:
     - 20121120_2012_PostSandy_NJ_000278_GeoClassified.las
     - 20121120_2012_PostSandy_NJ_000268_GeoClassified.las
   - Set the point spacing to **0.000001331**. The point spacing is obtained from calculating statistics or reading the lidar metadata.
   - Name the file **POST_ALL**, and save it to nj_results \NJ_Sandy.gdb\UTM\.

16. Open the Clip tool. Enter the following settings:
   - Set the input features to POST_ALL.
   - Set the clip features to study_area2.
   - Set the output feature class to \nj_results\NJ_Sandy.gdb\UTM**POST_ALL_SA2**.

17. Remove POST_ALL.

MAKING
SPATIAL
DECISIONS
USING GIS
AND LIDAR

SHORELINE
CHANGE AFTER
HURRICANE
SANDY

18. Search for and open the IDW tool (Spatial Analyst). Enter the following settings:
    - Set the input point features to POST_ALL_SA2.
    - Set the Z-value field to Shape.Z.
    - Set the output raster to nj_results**POST_ALL**. (Note: Do not save the raster to the geodatabase.)
    - Set the output cell size to **2**.

19. Remove POST_ALL_SA2.

20. For the POST_ALL raster, access the symbology properties and import the elevation layer file from the nj_results folder.

21. Turn POST_ALL on and off.

Q3   What is the biggest visual change?

The difference between the PRE and POST dataset is best highlighted by subtracting the PRE from the POST raster.

22. Open the Minus tool (Spatial Analyst). Enter the following settings:
    - Set the input raster or constant value 1 to POST_ALL.
    - Set the input raster or constant value 2 to PRE_ALL.
    - Set the output raster to \nj_results**Difference**.

23. Remove POST_ALL and PRE_ALL.

24. Open the symbology properties for the Difference raster. Enter the following settings:
    - In the Show panel, select the Stretched symbology.
    - Select the "Use hillshade effect" check box.
    - Set the high value to **9** and the low value to **–13**.

This coastline map will look better if it is rotated.

25. Right-click Layers, and go to Properties. On the General tab, set the rotation to **–75**.

26. Click the arrow next to the Add button, and select Add Data from ArcGIS Online. Search for Hurricane Sandy. Add the 201210 Hurricane Sandy USGS Imagery. Click noaa_01nov2012_utm18. mos, which gives you access to imagery along the damaged New Jersey coastline.

Q4 ***Where was the greatest elevation loss? Where was the greatest amount of erosion?***

Q5 ***Where was the greatest gain? Discuss the possible causes of this pattern.***

27. Create a layout using good cartographic principles. Save the map document as **nj_3.mxd**.

Deliverable 3: A map showing the difference in elevation between pre- and post-Sandy lidar data.

Draw conclusions and present the results

Once you have analyzed Hurricane Sandy's impact on Deal, New Jersey, including the change in shoreline elevation, choose a method of presenting your conclusions. Always keep the audience in mind as you prepare to report your results; they may not share your GIS expertise. Your results can be presented in a Word document, a PowerPoint presentation, or a more technical presentation mode, such as ArcGIS Online.

MAKING
SPATIAL
DECISIONS
USING GIS
AND LIDAR

2

SHORELINE
CHANGE AFTER
HURRICANE
SANDY

Pelham Manor, New York

Recommended deliverables

- **Deliverable 1:** A basemap of Pelham Manor, New York.
- **Deliverable 2:** A map containing two data frames showing pre- and post-Sandy lidar elevation.
- **Deliverable 3:** A map showing the difference in elevation between pre- and post-Sandy lidar data.

The questions asked in this project are both quantitative and qualitative. They identify key points that should be addressed in your analysis and final presentation.

Document your work, set environments, and examine the data

1. Open ArcMap. Add descriptive map document properties to your map document.

2. Set environments:
   - Open the data frame properties. Set the map projection to Geographic Coordinate Systems > World > WGS 1984.
   - On the Geoprocessing menu, click Environments.
   - Expand Workspace, and set the current workspace to \student\3_module8\ny_data.
   - Set the scratch workspace to \student\3_module8\ny_results.
   - For the output coordinate system, select Same as Display.

3. Save the map document as **ny_1** in the ny_results folder.

DATA (Current Workspace)   \student\3_module8\ny_data
RESULTS (Scratch Workspace)   \student\3_module8\ny_results

Analyze starting with a basemap

MAKING
SPATIAL
DECISIONS
USING GIS
AND LIDAR

SHORELINE
CHANGE AFTER
HURRICANE
SANDY

1. Add the westchester, study_area, and pelham_manor feature classes.

2. Add online imagery.

3. Create a locational map and save it as **ny_1**. Save the map document again as **ny_2**.

Q1 ***Describe the physical and demographic characteristics of Pelham Manor, New York.***

Deliverable 1: A basemap of Pelham Manor, New York.

Compare pre- and post-Sandy lidar data

1. Open ny_2.mxd. Add a data frame and name it **POST**. Change the name of the existing data frame to **PRE**.

2. In Catalog, create the PRE and POST LAS datasets. Use the data in the LAS_2012_PRE folder and LAS_2012_POST folders. Add these LAS datasets to the respective data frames.

3. Set the display properties to the maximum number of points.

4. Add the study_area feature class and zoom to the layer.

5. Symbolize the PRE and POST LAS datasets.

Q2 ***Describe any visible changes.***

6. Save the map document as **ny_2** to the ny_results folder.

Deliverable 2: A map containing two data frames showing pre- and post-storm lidar elevation.

MAKING
SPATIAL
DECISIONS
USING GIS
AND LIDAR

SHORELINE
CHANGE AFTER
HURRICANE
SANDY

Calculate the elevation difference between pre- and post-Sandy lidar data

1. Open a new ArcMap document and add document properties. Store relative pathnames.

2. Add the study_area feature class.

3. Set environments:
    *   Open the data frame properties. Set the map projection to NAD 1983 UTM Zone 18N.
    *   On the Geoprocessing menu, click Environments.
    *   Set the current workspace to \student\3_module8\ny_data.
    *   Set the scratch workspace to \student\3_module8\ny_results.
    *   For the output coordinate system, select Same as Display.
    *   Set the processing extent to Same as layer study_area.
    *   Set the raster analysis mask to study_area.

4. Save the map document as **ny_3.mxd** in the ny_results folder.

5. Open the LAS to Multipoint tool from LAS_2010_PRE. Enter the following settings:
    *   Set the point spacing to **0.0000066**. (The 0.0000066 point spacing was obtained from the metadata of the LAS dataset, or it can be accessed by calculating statistics on the LAS dataset.)
    *   Set study_area as soft clip.
    *   Save the feature class in the feature dataset in the NY.gdb.

6. Open the LAS to Multipoint tool from LAS_2012_POST. Set the point spacing to **0.000016**. Set study_area as soft clip.

7. Calculate the IDW using the IDW tool (Spatial Analyst), for both the pre- and post-Sandy data. Use cell sizes of **2** (meters) for the output rasters. Remember to store the output rasters in the results folder, not the geodatabase.

8. Using the Minus tool, perform a subtraction.

9. Create a map using good cartographic principles. Save it as **ny_3** in the ny_results folder.

Q3 **Where was the greatest elevation loss? Where was the most erosion?**

Q4 **Where was the greatest elevation gain? Discuss the possible causes of these patterns.**

Deliverable 3: A map showing the difference in elevation between pre- and post-Sandy lidar data.

Draw conclusions and present the results

Once you have analyzed Hurricane Sandy's effect on Pelham, New York, including the change in shoreline elevation, choose a method of presenting your conclusions. Always keep the audience in mind as you prepare to report your results; they may not share your GIS expertise. Your results can be presented in a Word document, a PowerPoint presentation, or a more technical presentation mode, such as ArcGIS Online.

MAKING
SPATIAL
DECISIONS
USING GIS
AND LIDAR

2

*SHORELINE
CHANGE AFTER
HURRICANE
SANDY*

MAKING
SPATIAL
DECISIONS
USING GIS
AND LIDAR

3

SHORELINE
CHANGE AFTER
HURRICANE
SANDY

PROJECT 3

On your own

Scenario

You have worked through a guided project and repeated the analysis for another project. In this project, you will reinforce your skills by researching and analyzing a similar scenario entirely on your own. First, identify your study area and acquire data for your analysis. You may want to study a local area.

Here is a list of topics that have been studied using lidar. The following represent possible project ideas:

- Forest characterization—canopy height and density
- Flood modeling
- Finding faults
- Geomorphic mapping
- Stream slope
- Archaeology field campaigns
- Mining—calculation of ore volumes
- Wind farm optimization

Many different websites distribute lidar data. Most of the available lidar data are within the United States. There are national data sites, and there are state data distributors. Reading the metadata and identifying a specific area of study is critical before downloading files. Some lidar files come compressed and require third-party software to convert them into the LAS format used by Esri.

- Open Topography facilitates community access to high-resolution, Earth science-oriented, topography data. It is a National Science Foundation-funded data facility. http://www.opentopography.org.
- The Earth Explorer provides online access to remotely sensed data from the US Geological Survey Earth Resources Observation and Science (EROS) Center archive. http://earthexplorer.usgs.gov/.
- National Oceanic and Atmospheric Administration, a world leader in coast science and management, provides state lidar datasets. http://www.csc.noaa.gov/dataviewer/#.
- Wikipedia lists national lidar datasets, organized according to state. http://en.wikipedia.org/wiki/National_LIDAR_Dataset_%E2%80%93_USA.

Research

Research the problem and answer the following questions:

1. What is the area of study?
2. What problem are you going to study?
3. What data is available?

MAKING
SPATIAL
DECISIONS
USING GIS
AND LIDAR

SHORELINE
CHANGE AFTER
HURRICANE
SANDY

MODULE 9

FOREST VEGETATION HEIGHT

SCENARIO ···

Land managers can learn a great deal about the history of a forested site based on the amount, distribution, and height of the vegetative cover. The effect of wildfire and disease, the growth of young trees, and the presence of habitat features favored by certain wildlife species are all important types of information that can be derived from lidar. Most of this information is currently collected through time-intensive ground surveys, often in remote locations. Increased efficiencies in data collection would be welcomed by land management agencies and advocacy organizations.

The Nature Conservancy's Virginia and Pennsylvania chapters want to use lidar to estimate the extent of various successional stages of forest evolution, using vegetation height as a surrogate for age. This knowledge will allow the land managers to better understand what restoration and management techniques may be necessary to maintain a diversity of forest communities and species. Lidar provides the opportunity to characterize different strata in ways that were previously not possible using satellite imagery. Canopy height can be determined by subtracting the bare earth surface (DEM) from the first return surface (DSM).

Projects included in this module

- **Project 1:** George Washington National Forest, Virginia
- **Project 2:** Michaux State Forest, Pennsylvania
- **Project 3:** On your own

Student worksheets

Project 1: George Washington National Forest, Virginia

- File name: 9a_VA_Forest_worksheet.docx
- Location: EsriPress\MSDLidar\student\4_module9\1va_documents

Project 2: Michaux State Forest, Pennsylvania

- File name: 9b_PA_Forest_worksheet.docx
- Location: EsriPress\MSDLidar\student\4_module9\2pa_documents

Objectives

- Investigate the LAS dataset for the designated forest area using profiles and calculating areas of profiles.
- Create DEM and DSM rasters from a LAS dataset.
- Use the Raster Calculator tool to subtract the DEM from the DSM to calculate canopy height.
- Identify and investigate problem areas in the lidar data.
- Produce a classified map of canopy vegetation.

PROJECT (1)

George Washington
National Forest, Virginia

The lidar data for this project was downloaded from the Virginia Lidar website:
http://virginialidar.com/.

Recommended deliverables

- **Deliverable 1:** A basemap showing the location of a LAS dataset.
- **Deliverable 2:** A map with two data frames showing a DEM and a DSM.
- **Deliverable 3:** A map showing canopy height.
- **Deliverable 4:** A graph of canopy height data.

The questions asked in this project are both quantitative and qualitative. They identify key
points that should be addressed in your analysis and final presentation.

Document your work, set environments,
and examine the data

1. Open ArcMap.

You need to add descriptive properties to every map document you produce. You can use the
same descriptive properties for every map document in the project or customize the documentation from map to map.

2. Add relevant map document properties. Store relative pathnames.

DATA (Current Workspace)   \student\4_module9\1va_data
RESULTS (Scratch Workspace)   \student\4_module9\1va_results

MAKING
SPATIAL
DECISIONS
USING GIS
AND LIDAR

FOREST
VEGETATION
HEIGHT

For this project, **1va_data** will be your project folder (EsriPress\MSDLidar\student\4_module9\1va_data). Make sure that it is stored in a place where you have write access. You can store your output data inside the **1va_results** folder (EsriPress\MSDLidar\student\4_module9\1va_results). The results folder contains an empty geodatabase named **va_forest_results.gdb** in which to save your data.

3. Save your map documents to the \student\6_module9\1va_results folder. Name the map document **va_canopy1**.

4. Set environments:
   - Open the data frame properties. Set the map projection to Projected Coordinate Systems > Lambert_Conformal_Conic_2SP. Lambert Conformal Conic 2SP is a custom projection used in Virginia. This projection can be added by importing it from the layers feature dataset within the va_forest.gdb.
   - On the Geoprocessing menu, click Environments.
   - Expand Workspace, and set the current workspace to \student\4_module9\1va_data.
   - Set the scratch workspace to \student\4_module9\1va_results.
   - For the output coordinate system, select Same as Display.

Analyze starting with a basemap

1. Connect to the \student\4_module9\1va_data folder.

In this folder, you will see another folder that holds the LAS files for the George Washington National Forest in Virginia. The metadata for the LAS files is as follows:

- VA_Augusta_2011
- Lambert_Conformal_Conic_2SP
- NAVD_1988_Feet
- Load Data 10/2011
- Tile 1.5 × 1.5 square mile
- 2011_VA_Augusta_2011_n16_3803_20
- 2011_VA_Augusta_2011_n16_3803_10

2. From the va_forest geodatabase, add the va_counties, GWNF, and va_forest.lasd layers.

Q1 **Do the coordinate systems for the feature dataset match the coordinate system of the LAS files?**

DATA (Current Workspace)   \student\4_module9\1va_data
RESULTS (Scratch Workspace)   \student\4_module9\1va_results

3. Add the Imagery with Labels basemap.

Q2 **Describe the landscape surrounding the LAS data frame.**

4. Remove the Imagery with Labels basemap and the References layer. Add the Topographic basemap.

5. Remove the basemap.

6. Switch to layout view. Add all essential map elements (scale, north arrow, title, and legend). Label the forest and the counties.

7. Save the map document as **va_canopy1**.

8. Save the file again as **va_canopy2**.

Deliverable 1: A basemap showing the location of a LAS dataset.

Lidar data can be classified into various heights by selecting the proper codes. Remember, a DEM (digital elevation model) is a bare-earth model that uses the last return. This can be compared to a DSM (digital surface model) created from first returns or from the highest points above the ground.

The metadata for the LAS files is in \student\4_module9\1va_data\LAS_files\FGDC_USGS_NRCS_VA_LAS.xml. In the metadata, the classification scheme is given as follows:
- Class 1 = Unclassified. This class includes vegetation, buildings, noise, and so on.
- Class 2 = Ground.

Using this information, we can filter the LAS data so that Class 1 gives the points for the DSM, and Class 2 gives the data for the DEM. This class data can also be accessed under the Filters menu on the LAS Dataset toolbar. The metadata gives the point spacing as 1.53 and 1.60, or an average of 1.57 feet.

Raster data is one of the most common GIS data types. A wide range of analysis can be done with raster or gridded data. For the vegetation height analysis, you will convert the LAS dataset into a DEM and a DSM.

MAKING
SPATIAL
DECISIONS
USING GIS
AND LIDAR

FOREST
VEGETATION
HEIGHT

Analyze: Create a DEM

1. Open va_canopy2. Switch to data view.

2. Remove the va_counties and GWNF data layers.

3. Insert a new data frame and name it **DEM**. Copy the va_forsest.las dataset from the Layers data frame into the DEM data frame.

4. Right-click va_forest.lasd, and go to Properties. On the Display tab, enter the following settings:
   - Set the point limit to **5000000**.
   - Select the "Use scale to control full resolution" check box.
   - Set the full resolution scale to **5000**.
   - Move the Point Density slider to Fine.

5. Turn on the LAS Dataset toolbar. From the Filters menu, select Ground.

6. Open the LAS Dataset to Raster tool. Enter the following settings:
   - Set the input LAS dataset to va_forest.lasd.
   - Set the output raster to \student\4_module9\1va_results\va_forest_results.gdb**DEM**.
   - Set the value field to ELEVATION.
   - Binning represents the interpolation method used to produce the raster.
     - Set the cell assignment type to MAXIMUM.
     - Set the void fill method to NATURAL_NEIGHBOR.
   - Set the output data type to INT (for integer).
   - Set the sampling type to CELLSIZE.
   - Set the sampling value to **6**. (This value is used to define the resolution of the output raster.)
   - Accept the remaining default values.
   - Click OK.

Q3 ***What are the lowest and highest elevations represented in the DEM?***

7. Right-click DEM, and click Properties. On the Symbology tab, pick an appropriate elevation color ramp.

8. From the basemaps, add the Terrain with Labels basemap.

DATA (Current Workspace)   \student\4_module9\1va_data
RESULTS (Scratch Workspace)   \student\4_module9\1va_results

Q4   ***Write a brief spatial description of the data frame.***

9.   Remove the Terrain basemap and the References layer.

Analyze: Create a DSM

1.   Insert a new data frame and name it **DSM**.

2.   Copy the va_forest.lasd dataset from the DEM data frame and paste it into the DSM data frame.

3.   Turn on the LAS Dataset toolbar. From the Filters menu, select Non Ground.

4.   Open the LAS Dataset to Raster tool and enter the same parameters used in the previous section to create the DEM. Name the output raster **DSM**.

5.   Switch to layout view. Construct a layout containing two data frames showing the DEM and the DSM. Use appropriate map elements and cartographic principles.

6.   Save the map as **va_canopy2** and again as **va_canopy3**.

Deliverable 2: A map with two data frames showing a DEM and a DSM.

Calculate the vegetation height and investigate areas with negative height values

To determine the vegetation height, the bare earth surface (DEM) is subtracted from the DSM or first return.

1.   Switch to data view.

You need to get DEM_ft and DSM_ft in the same data frame.

2.   Rename the DEM data frame **Vegetation Height**. Copy the DSM data layer from the DSM data frame and paste it into the Vegetation Height data frame. Remove the DSM data frame from the map document.

MAKING
SPATIAL
DECISIONS
USING GIS
AND LIDAR

FOREST
VEGETATION
HEIGHT

3. Search for and open the Minus tool (Spatial Analyst). Enter the following settings:
- Input DSM as constant value 1.
- Input DEM as constant value 2.
- Name the output raster **height** and save it to the va_forest_results.gdb.

Q5 ***What are the lowest and highest values in the height raster?***

Obviously, the negative height values indicate errors. Any heights over 196 feet are also errors. There are no manufactured structures in the study area, and no trees in Virginia are over 196 feet tall. The errors in the data are probably a misclassification of the ground points.

4. Open the height raster attribute table and investigate the data.

Q6 ***How many cells have a value over 196 feet?***

Q7 ***How many cells have a value less than 0?***

The eleven cells with a value over 196 feet are insignificant. However, you should investigate the 11,719 cells with negative values and search for possible reasons.

5. To investigate the cells with values less than 0, you must first isolate them. Run the Raster Calculator tool with a Map Algebra expression of "height"< 0. Name the output raster **negative** and save it to the va_forest_results.gdb.

Q8 ***Describe the distribution of the cells that are less than 0.***

6. Click the arrow next to the Add button and select Add Data from ArcGIS Online.

7. Search for **Virginia NAIP 2012 1 m**.

8. Add the Virginia NAIP imagery and examine the three problem areas.

9. Make the 0 value of the negative raster hollow. Choose a bright color that will contrast with the air photo's forest color so that you see only the negative-valued cells.

10. Zoom to the three problem areas.

DATA (Current Workspace)   \student\4_module9\1va_data
RESULTS (Scratch Workspace)   \student\4_module9\1va_results

Q9 *Describe the imagery of the problem areas.*

MAKING
SPATIAL
DECISIONS
USING GIS
AND LIDAR

*FOREST
VEGETATION
HEIGHT*

Classify the vegetation height

1. Remove the NAIP imagery, LAS dataset, negative, DSM, and DEM layers from the map.

You need to classify the height raster, leaving out the negative cells and the cells over 196 feet.

2. Right-click height and go to Properties. On the Symbology tab, in the Show panel, click Categories. Select Unique Values.

3. Right-click "values less than or equal to 0," and click Remove Value(s).

4. Select 1–5. Right-click your selection and click Group Values. Name the new group **shrub**. Change the color of the new group to light yellow.

5. Repeat the process in step 4 for values 6–15. Name the new group **Small Regen** (regenerative growth). Change the color to light green.

MAKING
SPATIAL
DECISIONS
USING GIS
AND LIDAR

FOREST
VEGETATION
HEIGHT

6. Repeat the process in step 4 for values 16–25. Name the new group **Large Regen**. Change the color to medium green.

7. Repeat the process in step 4 for values 26+. Name the new group **Trees**. Change the color to dark green.

8. Remove values greater than 196.

Q10 *Can you see any man-made structures? Describe them.*

Q11 *Does the man-made structure seem to have any influence on the vegetation?*

9. Switch to layout view. Construct a layout showing vegetation height. Use appropriate map elements and cartographic principles.

10. Save your map as **va_canopy3**.

Deliverable 3: A map showing canopy height.

Graph tree height data

Another way of viewing the tree height data is to create a bar graph showing the height of trees by the number of cells.

1. Go to View > Graphs > Create Graph. Enter the following settings:
- Set the value field to count.
- Clear the "Add to legend" check box.
- Set the color to Match with Layer.
- Click Next.
- Type the title **Height of Vegetation**.
- Select the "Graph in 3D view" check box.
- In the Axis Properties, click the Bottom tab. Type the title **Height in ft**.
- Click Finish.

DATA (Current Workspace)   \student\4_module9\1va_data
RESULTS (Scratch Workspace)   \student\4_module9\1va_results

MAKING
SPATIAL
DECISIONS
USING GIS
AND LIDAR

FOREST
VEGETATION
HEIGHT

Q12 *What does each point represent?*

Q13 *What does the graph tell you about the forest in the designated study area?*

Deliverable 4: A graph of canopy height data.

Draw profiles

A transect is a line that spans an area of interest and allows the analyst to locate sample objects. The transect length is determined by the study area and the problem. Transects are used in forestry to estimate the proportion of different trees in a study area. Transects are accurate enough for some types of forest management, including infected tree population identification and forest-edge analysis.

A common way to visualize lidar data is to use a 2D cross-sectional view. A selected set of lidar points from a LAS dataset can be displayed using the Profile View window accessed from the LAS Dataset toolbar.

To do a detailed analysis of transects, perform the following steps:

1. Add the va_forest.lasd dataset back to the table of contents.

2. Use the Profile tool to draw transect lines across areas that you want to more closely investigate.

Draw conclusions and present the results

After analyzing the distribution of canopy heights in the study area, choose a method of presenting your conclusions. Always keep the audience in mind as you prepare to report your results; they may not share your GIS expertise. Your results can be presented in a Word document, a PowerPoint presentation, or a more technical presentation mode, such as ArcGIS Online.

MAKING
SPATIAL
DECISIONS
USING GIS
AND LIDAR

2

FOREST
VEGETATION
HEIGHT

PROJECT 2

Michaux State Forest, Pennsylvania

The lidar data for this project was downloaded from PASDA (Pennsylvania Spatial Data Access): http://www.pasda.psu.edu/default.asp.

Recommended deliverables

- **Deliverable 1:** A basemap showing the location of a LAS dataset.
- **Deliverable 2:** A map with two data frames showing a DEM and a DSM.
- **Deliverable 3:** A map showing canopy height.
- **Deliverable 4:** A graph of canopy height data.

The questions asked in this project are both quantitative and qualitative. They identify key points that should be addressed in your analysis and final presentation.

Document your work, set environments, and examine the data

You need to add descriptive properties to every map document you produce. You can use the same descriptive properties for every map document in the project or customize the documentation from map to map.

1. Add relevant document properties. Store relative pathnames.

For this project, **2pa_data** will be your project folder (EsriPress\MSDLidar\student\4_module9\2pa_data). Make sure that it is stored in a place where you have write access. You can store your output data inside the **2pa_results** folder (EsriPress\MSDLidar\student\4_module9\2pa_results). The results folder contains an empty geodatabase named **pa_forest_results.gdb** in which to save your data.

DATA (Current Workspace)   \student\4_module9\2pa_data
RESULTS (Scratch Workspace)   \student\4_module9\2pa_results

2. Save your map documents to the student\4_module9\2pa_results folder. Name the file **pa_canopy1**.

3. Set environments:
- Open the data frame properties. Set the map projection to Projected Coordinate Systems > State Plane > NAD 1983 (US Feet) > NAD_1983_StatePlane_Pennsylvania_South_FIPS_3702_Feet.
- On the Geoprocessing menu, click Environments.
- Expand Workspace, and set the current workspace to \student\4_module9\2_pa_data.
- Set the scratch workspace to \student\4_module9\2pa_results.
- For the output coordinate system, select Same as Display.

MAKING
SPATIAL
DECISIONS
USING GIS
AND LIDAR

FOREST
VEGETATION
HEIGHT

Analyze starting with a basemap

1. Connect to the \student\4_module9\2pa_data folder.

In this folder, you will see another folder that holds the LAS files for the Michaux State Forest, Pennsylvania. The following is the metadata for the LAS files:
- PAMAP Program, PA Department of Conservation and Natural Resources, Bureau of Topographic and Geologic Survey
- NAD_1983_StatePlane_Pennsylvania_South_FIPS_3702_Feet
- NAVD_1988_Feet
- Load Data 04/2006
- Tile 1.9 × 1.9 square mile
- Filename: 24002060PAS
- Filename: 24002070PAS

2. Add the following layers: pa_counties, pa_forest.lasd, places, michaux_sf, and pa_breaklines.

Q1 **Do the coordinate systems for the feature dataset match the coordinate system of the LAS files?**

3. Add the Topographic basemap.

Q2 **In what range of mountains is the LAS data frame?**

Q3 **What do the pa_breaklines represent? (Hint: Read the metadata.)**

4. Remove the basemap.

MAKING
SPATIAL
DECISIONS
USING GIS
AND LIDAR

FOREST
VEGETATION
HEIGHT

5. Switch to layout view. Add all essential map elements (scale, north arrow, title, and legend). Label the forest and counties.

6. Save the file as **pa_canopy1**.

7. Save the file again as **pa_canopy2**.

Deliverable 1: A basemap showing the location of a LAS dataset.

The metadata for the LAS files is in student\4_module9\2pa_data\LAS_files\pamap_Lidar_LAS.xml. In the metadata, the classification is defined as follows:

- Class 1 = Bridges, overpasses, buildings, cars, and vegetation
- Class 2 = Ground
- Class 9 = Water
- Class 12 = Non Ground, most likely vegetation

Using this information, we can filter the LAS data so that Classes 1 and 12 give the points for the DSM, and Class 2 gives the data for the DEM. This class data can also be accessed under the Filters menu on the LAS Dataset toolbar. The metadata also tells us that the average point spacing was 1.4.

Analyze: Create a DEM

1. Calculate the DEM of PA_LAS_Dataset. Refer to the directions in project 1, if necessary.

2. Use the LAS Dataset to Raster tool. Enter the following settings:
- Set the input LAS dataset to pa_forest.lasd.
- Set the output raster to \student\4_module9\2pa_results\pa_forest_results.gdb**DEM**.
- Set the value field to ELEVATION.
- Binning represents the interpolation method used to produce the raster.
  - Set the cell assignment type to MAXIMUM.
  - Set the void fill method to NATURAL_NEIGHBOR.
- Set the output data type to INT (for integer).
- Set the sampling type to CELLSIZE.
- Set the sampling value to **6**.
- Accept the remaining default values.
- Click OK.

DATA (Current Workspace)   \student\4_module9\2pa_data
RESULTS (Scratch Workspace)   \student\4_module9\2pa_results

3. Rename the data frame **DEM**.

Q4 ***What are the lowest and highest elevations represented in the DEM?***

Q5 ***Write a brief description of the DEM.***

Analyze: Create a DSM

1. Insert a new data frame. Name the new data frame **DSM**.

2. Calculate the DSM.

Remember: Set the filter to Non Ground.

Deliverable 2: A map with two data frames showing a DEM and a DSM.

Calculate the vegetation height and investigate areas with negative height values

To determine the vegetation height, the bare earth surface (DEM) is subtracted from the DSM or first return. Use the Minus tool to create a height raster.

Q6 ***What are the lowest and highest values in the height raster?***

Q7 ***How many cells have a value over 196 feet?***

Q8 ***How many cells have a value less than 0?***

Q9 ***Describe the distribution of the cells that are less than 0.***

Classify the vegetation height

1. Classify the height raster leaving out the negative cells and the cells over 196 feet. Use the following information to classify the vegetation height map:
- 0–5 feet = Shrub
- 6–15 feet = Small Regen
- 16–25 feet = Large Regen
- >25 feet = Trees

MAKING
SPATIAL
DECISIONS
USING GIS
AND LIDAR

FOREST
VEGETATION
HEIGHT

MAKING
SPATIAL
DECISIONS
USING GIS
AND LIDAR

FOREST
VEGETATION
HEIGHT

Q10 *Can you see any man-made structures? Describe them.*

2. Switch to layout view and construct a layout showing vegetation height. Use appropriate map elements and cartographic principles.

3. Save as **pa_canopy3**.

Deliverable 3: A map showing canopy height.

Graph tree height data

1. Create a graph of the vegetation height.

Q11 *What does the graph tell you about the forest in the designated study area?*

Deliverable 4: A graph of canopy height data.

Draw profiles

To do a detailed analysis of transects, perform the following steps:

1. Add the pa_forest.lasd dataset back to the table of contents.

2. Use the Profile tool to draw transect lines across areas that you want to more closely investigate.

3. Save your map document and exit ArcMap.

Draw conclusions and present the results

Once you have completed your analysis of the variation in canopy heights, choose a method of presenting your conclusions. Always keep the audience in mind as you prepare to report your results; they may not share your GIS expertise. Your results can be presented in a Word document, a PowerPoint presentation, or a more technical presentation mode, such as ArcGIS Online.

DATA (Current Workspace)   \student\4_module9\2pa_data
RESULTS (Scratch Workspace)   \student\4_module9\2pa_results

PROJECT 3

On your own

MAKING
SPATIAL
DECISIONS
USING GIS
AND LIDAR

FOREST
VEGETATION
HEIGHT

Scenario

You have worked through a guided project and repeated the analysis for another project. In this project, you will reinforce your skills by researching and analyzing a similar scenario entirely on your own. First, identify your study area and acquire data for your analysis. You may want to study a local area.

Here is a list of topics that have been studied using lidar. The following represent possible project ideas:

- Forest characterization—canopy height and density
- Flood modeling
- Finding faults
- Geomorphic mapping
- Stream slope
- Archaeology field campaigns
- Mining—calculation of ore volumes
- Wind farm optimization

MAKING
SPATIAL
DECISIONS
USING GIS
AND LIDAR

3

*FOREST
VEGETATION
HEIGHT*

Many different websites distribute lidar data. Most of the available lidar data are within the United States. There are national data sites, and there are state data distributors. Reading the metadata and identifying a specific area of study is critical before downloading files. Some lidar files come compressed and require third-party software to convert them into the LAS format used by Esri.

- Open Topography facilitates community access to high-resolution, Earth science-oriented, topography data. It is a National Science Foundation-funded data facility. http://www.opentopography.org.
- The Earth Explorer provides online access to remotely sensed data from the US Geological Survey Earth Resources Observation and Science (EROS) Center archive. http://earthexplorer.usgs.gov/.
- National Oceanic and Atmospheric Administration, a world leader in coast science and management, provides state lidar datasets. http://www.csc.noaa.gov/dataviewer/#.
- Wikipedia lists national lidar datasets, organized according to state. http://en.wikipedia.org/wiki/National_LIDAR_Dataset_%E2%80%93_USA.

Research

Research the problem and answer the following questions:

1. What is the area of study?
2. What problem are you going to study?
3. What data is available?

DEPRESSIONAL WETLAND DELINEATION FROM LIDAR

SCENARIO ···

Florida has a high percentage of wetlands. Wetlands have been protected in the United States since the 1970s under the Federal Clean Water Act. Once wetlands are identified, they can be declared subject to the Clean Water Act of 1977 and protected. These depressed topographical areas can collect and store water as well as provide a rich habitat for many plants and animals.

Wetlands have traditionally been delineated using field inventories. Such inventories require considerable time in the field by experienced scientists to make observations. Lidar, combined with the use of GIS, offers an opportunity for off-site delineation of wetlands.

The State of Florida has decided to establish a wetland center to do advanced research comparing the traditional field-based delineation of wetlands to off-site delineation using lidar. The State has asked for an analysis exploring the use of lidar to delineate depressional wetlands. Lidar is well suited for the detection of depressional wetlands because of the strong relationship between the wetland boundary and elevation. They have asked for a comprehensive comparison between field-based and off-site lidar-based wetland delineations using multiple methods of analysis. For the purpose of this study, wetlands identified by the National Wetlands Inventory (NWI) will be used as a surrogate for the field-based component.

Projects included in this module

- **Project 1:** Wakulla, Florida
- **Project 2:** Pasco County, Florida
- **Project 3:** On your own

Student worksheets

Project 1: Wakulla, Florida

- File name: 10a_wakulla_worksheet.docx
- Location: EsriPress\MSDLidar\student\5_module10\1fl_documents

Project 2: Pasco County, Florida

- File name: 10b_pasco_worksheet.docx
- Location: EsriPress\MSDLidar\student\5_module10\2fl_documents

Objectives

- Produce lidar TINs with contours.
- Produce lidar terrains with contours.
- Convert DEM point cloud to raster by kriging with contours.
- Compare the three types of wetland delineation to the NWI delineation.

PROJECT 1

Wakulla, Florida

The lidar data for this project is from Florida International University:
http://digir.fiu.edu/Lidar/LidarNew.php.

Wetlands data is from the US Fish and Wildlife Service:
http://www.fws.gov/wetlands/Data/Data-Download.html.

Recommended deliverables

- **Deliverable 1:** A basemap showing wetland types and LAS tiles.
- **Deliverable 2:** A map showing three types of wetland delineation. The map should have four data frames: TIN, Terrain, Kriging, and Comparison.

The questions asked in this project are both quantitative and qualitative. They identify key points that should be addressed in your analysis and final presentation.

Document your work, set environments, and examine the data

1. Open ArcMap. Add descriptive map document properties. Store relative pathnames.

For this project, **1fl_data** will be your project folder (EsriPress\MSDLidar\student \5_module10\1fl_data). Make sure that it is stored in a place where you have write access. You can store your output data inside the **1fl_results** folder (EsriPress\MSDLidar \student\5_module10\1fl_results). The results folder contains an empty geodatabase named **fl_wetlands_results.gdb** in which to save your data.

MAKING
SPATIAL
DECISIONS
USING GIS
AND LIDAR

1

DEPRESSIONAL
WETLAND
DELINEATION
FROM LIDAR

2. Save your map documents inside the \student\5_module10\1fl_results folder. Name the map document **flwetlands1**.

3. Set environments:
   - Open the data frame properties. Set the map projection to Custom > Lambert_Conformal_Conic_2SP.
   - On the Geoprocessing menu, click Environments.
   - Expand Workspace, and set the current workspace to \student\5_module10\1fl_data.
   - Set the scratch workspace to \student\5_module10\1fl_results.
   - For the output coordinate system, select Same as Display.

Analyze starting with a basemap

Construct a basemap

1. Connect to the folder \student\5_module10\1fl_data.

In this folder, you will see another folder that holds the LAS files for Wakulla County, Florida. The following is the metadata for the LAS files.
- County and Date: Wakulla_County_2008
- Coordinate System: Lambert_Conformal_Conic_2SP
- Datum: NAVD_1988_Feet
- Load Data: 2007
- Tile size: 1 × 1 square mile
- Filename: LID_2007_065029_N.las
- Filename: LID_2007_065031_N.las

2. Add the following layers to the map document: counties, wakulla, wetlands, and tile_29.lasd.

3. Open the wetlands attribute table and look at the different types of wetlands.

Q1 *List the different types of wetlands.*

4. Before zooming to the study area, add the Imagery basemap.

5. Set the counties data layer symbology to a county boundary outline.

6. Set the wakulla data layer symbology to a hollow outline.

MAKING
SPATIAL
DECISIONS
USING GIS
AND LIDAR

1

DEPRESSIONAL
WETLAND
DELINEATION
FROM LIDAR

Q2 ***Describe Wakulla County.***

7. Zoom to the area designated by the wetlands feature class. Classify the wetlands by WETLAND_TYPE and make them an appropriate color.

Q3 ***What are the lowest and highest elevations shown in the LAS dataset?***

8. Create a basemap of the area using correct cartographic principles. Save the map document as **flwetlands1** to your 1fl_results folder. Use the online imagery basemap. Save the map again as **flwetlandstile29**.

Deliverable 1: A basemap showing wetland types and LAS tiles.

Derive TINs and contours from LAS datasets

The first method of wetland delineation is converting the LAS dataset into a bare earth TIN. Each tile must be done separately.

1. Open flwetlandstile29 and remove all data layers except tile29.lasd.

2. Add the tile29_wetlands_sa feature class. Symbolize it with a hollow outline with a wide border. Label the site using the WETLAND_TYPE field.

3. Name the data frame **LAS Dataset to TIN**.

4. Select tile29.lasd in the LAS Dataset toolbar.

5. Set the filter to Ground.

6. Search for and open the LAS Dataset to TIN tool. Enter the following settings:
   - Set the input LAS dataset to tile_29.lasd.
   - Set the output TIN to student\5_modules10\1fl_ results**tile29_TIN**.
   - Set the thinning type to RANDOM.
   - Set the thinning method to PERCENT.
   - Set the thinning value to **75**.
   - Set the maximum value to **8000000**.
   - Z Factor = **1**.
   - Click OK.

MAKING
SPATIAL
DECISIONS
USING GIS
AND LIDAR

1

DEPRESSIONAL
WETLAND
DELINEATION
FROM LIDAR

Q4 ***What are the lowest and highest elevations shown in the tile29_TIN?***

7. Remove tile_29.lasd.

8. Right-click tile29_TIN and go to Properties. On the Symbology tab, classify the data into 5 classes. Use the Defined Interval classification method. Choose appropriate break values. Save your map document.

The next part of the exercise uses the Surface Contour tool. This tool uses a triangulated irregular network (TIN) or a terrain dataset to calculate contours. Contours are generated directly from the TIN or terrain dataset within its zone of interpolation using linear interpolation. With this interpolation, each triangle is treated as a plane. Portions of individual contours within a triangle are straight. Any change in direction occurs only when a contour passes from one triangle into another. This type of contouring produces engineering-quality contours, representing an exact linear interpolation of the surface model.

Linear interpolation is generally considered conservative and often represents the best estimate for analysis. The resulting contours are not smooth, though, and generally are not used for aesthetic cartographic output.

9. Search for and open the Surface Contour tool. Enter the following settings:
   - Set the input TIN to tile29_TIN.
   - Set the output feature class to fl_wetlands_result.gdb\Layers**TINcontour_1ft**.
   - Set the contour interval to **1**.
   - Accept the remaining default parameter values.
   - Click OK.

10. Zoom to the tile29_wetlands_sa data layer.

11. Right-click TINcontour_1ft and go to Selection > Make This The Only Selectable Layer. Select the contour that most closely matches tile29_wetlands_sa.

12. Right-click TINcontour_1ft. Go to Data > Export Data. Save the selected feature to \fl_wetlands_results.gdb\Layers. Name the new feature class **sa_TIN_1ft**.

13. Remove TINcontour_1ft.

DATA (Current Workspace)   \student\5_module10\1fl_data
RESULTS (Scratch Workspace)   \student\5_module10\1fl_results

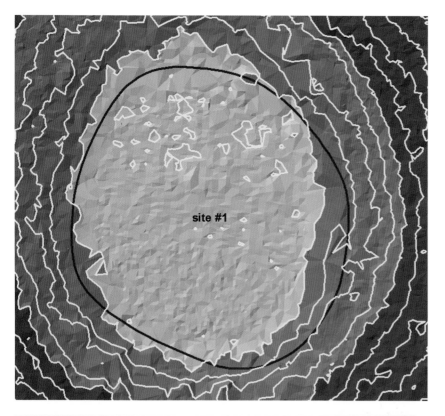

DEPRESSIONAL
WETLAND
DELINEATION
FROM LIDAR

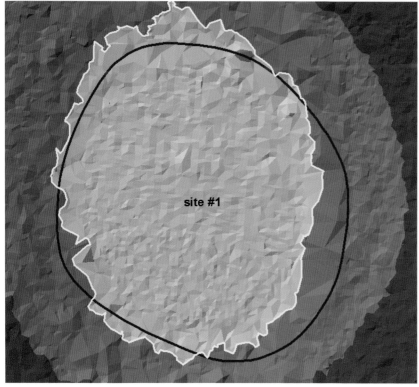

DATA (Current Workspace)   \student\5_module10\1fl_data
RESULTS (Scratch Workspace)   \student\5_module10\1fl_results

MAKING
SPATIAL
DECISIONS
USING GIS
AND LIDAR

1

DEPRESSIONAL
WETLAND
DELINEATION
FROM LIDAR

You need to convert the line feature to a polygon feature to compare the acreage for the derived TIN contour.

14. Search for and open the Feature to Polygon tool.
    • Set the input features to sa_TIN_1ft.
    • Set the output feature class to fl_wetlands_result.gdb\Layers**poly_TIN_29**.
    • Click OK.

15. Make the polygon hollow. Set the polygon border width to **3** and make it a distinctive color.

16. Remove sa_TIN_1ft.

If you open the tile29_wetlands_sa attribute table, you will see that the area is not only given in square feet (in the SHAPE_Area field) but also in acres. It would be much easier to make comparisons if poly_TIN_29 also had area units in acres. One acre is equal to 43,560 square feet.

17. Open the attribute table for poly_TIN_29. In the upper left, from the Table options menu, select Add Field. Enter the following settings:
    • For the name, type **Acres**.
    • Set the type to float.

18. After the Acres field is created, right-click it and go to Field Calculator. Click Yes on the warning about calculating outside the edit session. Type **[Shape_Area] /43560**.

You should now have two delineated wetland polygons for site 1. The original polygon was taken from the National Wetlands Inventory, and the second polygon was calculated by changing a lidar LAS dataset to a TIN. This completes the first type of wetland delineation.

Derive contours from terrains

Terrains are TIN-based surfaces built from measurements and are stored as features in a geodatabase. They are useful in managing lidar because of the massively large point dataset elevations. Terrain pyramids are levels of detail generated for a terrain dataset to improve efficiency in rendering the terrain on the screen. They are used as a form of scale-dependent generalization. Pyramid levels take advantage of the fact that accuracy requirements diminish with scale.

Terrain pyramids are generated through the process of point reduction, also known as point thinning. This reduces the number of measurements needed to represent a surface for a given area. For each successive pyramid level, fewer measurements are used, and the accuracy requirements necessary to display the surface drop accordingly. The original source measurements are still used in coarser pyramids, but there are fewer of them. No resampling, averaging, or derivative data are used for pyramids.

1. Insert a new data frame and name it **Terrain**.

2. Copy the tile29_wetlands_sa feature class and paste it into the new Terrain data frame.

3. Open the LAS to Multipoint tool. Enter the following settings:
   * Set the input to LID2007_065029_N.las from the LAS_files folder in 1fl_data.
   * Set the output feature class to fl_wetlands_result.gdb\Layers**tile29_ptcloud_dem**.
   * Set the average point spacing to **1.6**.
   * Set the input class code to **2** (Ground).
   * Accept the remaining default parameters.
   * Click OK.

4. In the fl_wetlands_results.gdb, right-click Layers and go to New > Terrain. Enter the following settings:
   * Name the terrain **tile29terrain**.
   * Select the tile29ptcloud_dem.
   * For approximate point spacing, type **1.6**.
   * Click Next.
   * Click Next again.
   * On the Select pyramid type screen, for window size point selection method, choose Z Mean. Window size-based thinning is fastest and is good at reducing noise. Leave the secondary thinning method set to None.
   * Click Next.
   * Click Calculate Pyramid Properties, and then add two more pyramids (numbers 4 and 5).
   * Click Next, Finish, and then Yes to create the new terrain.

Add the tile29terrain to the Terrain data frame.

5. Right-click tile29terrain and go to Properties > Symbology.

6. Highlight Elevation on the left side of the Symbology dialog box. Classify using 5 classes. Set the classification method to Defined Interval and choose appropriate break values.

7. Remove tile29_ptcloud_dem.

The Surface Contour tool uses an input terrain dataset to calculate contours. The output is a 2D polyline feature class with contour heights assigned as attributes.

MAKING
SPATIAL
DECISIONS
USING GIS
AND LIDAR

1

DEPRESSIONAL
WETLAND
DELINEATION
FROM LIDAR

8. Search for and open the Surface Contour tool. Enter the following settings:
   - Set the input surface to tile29terrain.
   - Set the output feature class to fl_wetlands_results.gdb\Layers**terrain29_1ft**.
   - Set the contour interval to **1**.
   - Accept the remaining default parameters.
   - Click OK.

9. Zoom in. From the new terrain29_1ft feature class, select the contour that is closest to the site 1 polygon (the tile29_wetlands_sa feature class).

10. Search for and open the Feature to Polygon tool. Enter the following settings:
    - Set the input features to terrain29_1ft.
    - Set the output feature class to fl_wetlands_results.gdb\Layers**poly_terrain_29**.
    - Accept the remaining default parameters.
    - Click OK.

11. Make poly_terrain_29 hollow. Increase the border width, and make the border a distinct color.

12. Remove terrain29_1ft.

Calculate acreage for poly_terrain_29 by performing the following steps:

13. Open the poly_terrain_29 attribute table. From the Table Options menu, choose Add Field. Enter the following settings:
    - For the name, type **Acres**.
    - Set the type to float.

14. After creating the Acres field, right-click it and choose Calculate Geometry. Calculate Acres US [ac].

Derive contours from kriging

Kriging is an advanced geostatistical procedure that generates an estimated surface from a scattered set of points with z-values. Unlike other interpolation methods, the Kriging tool effectively involves an interactive investigation of the spatial behavior of the phenomenon represented by the z-values before you select the best estimation method for generating the output surface.

1. Insert a new data frame and name it **Kriging**.

2. Copy the tile29_wetlands_sa data layer and paste it into the Kriging data frame.

DATA (Current Workspace)   \student\5_module10\1fl_data
RESULTS (Scratch Workspace)   \student\5_module10\1fl_results

3. From \1fl_data\fl_wetlands.gdb\Layers, add sa_kriging. From \1fl_results\fl_wetlands_results .gdb\Layers, add tile29_ptcloud.

4. On the Geoprocessing menu, click Clip to open the Clip tool. Enter the following settings:
   - Set the input feature to tile29_ptcloud_dem.
   - Set the clip feature to sa_kriging.
   - Set the output feature class to fl_wetlands_results.gdb\Layers**sa_kriging**.

5. Remove sa_kriging and tile29_ptcloud_dem from the Kriging data frame.

6. Search for and open the Kriging tool. Enter the following settings:
   - Set the input sa_kriging.
   - Set the Z-value field to Shape.Z.
   - Set the output surface raster to tile_29kriging. (Note: Store the output surface raster in fl_ wetlands_results.gdb. The raster cannot be stored in the Layers feature dataset.)
   - Accept the remaining defaults.
   - Click OK.

7. Remove the sa_kriging feature class.

8. Right-click tile29_kriging and go to Properties > Symbology. In the Show panel, choose Classify. Classify the data into 5 classes. Set the classification method to Defined Interval and choose appropriate break values.

9. Search for and open the Contour (Spatial Analyst) tool. Make sure not to use the Surface Contour (3D Analyst) tool. Enter the following settings:
   - Set the input raster to tile29_kriging.
   - Set the output feature class to fl_wetlands_results.gdb\Layers**kriging29_1ft**.
   - Set the contour interval to **1**.
   - Accept the remaining default parameters.
   - Click OK.

10. Zoom to the tile29_wetlands_sa feature class. Click the Select Features button to select the kriging29_1ft contour that lies closest to site 1 polygon.

11. Search for and open the Feature to Polygon tool. Enter the following settings:
    - Set the input features to kriging29_1ft.
    - Set the output feature class to fl_wetlands_results.gdb\Layers**poly_kriging_29**.
    - Click OK.

DATA (Current Workspace) \student\5_module10\1fl_data
RESULTS (Scratch Workspace) \student\5_module10\1fl_results

MAKING
SPATIAL
DECISIONS
USING GIS
AND LIDAR

DEPRESSIONAL
WETLAND
DELINEATION
FROM LIDAR

12. Make poly_kriging_29 hollow. Increase the border width, and make the border a distinct color.

13. Remove kriging29_1ft.

14. Open the poly_kriging_29 attribute table. From the Table Options menu, click Add Field. Enter the following settings:
- For the name, type **Acres**.
- Set the type to float.

15. After you create the Acres field, right-click it and choose Calculate Geometry. Calculate Acres US [ac].

Compare wetland delineations

1. Insert a fourth data frame into your map document. Name it **Wetland Comparison**.

2. Add the tile29_wetlands_sa, poly_TIN_29, poly_terrain_29, and poly_kriging_29 feature classes. Make all the data layers hollow and choose distinct border colors for each.

Q5 **Compare the contours visually. A good way to do this is to create a chart similar to the one on your worksheet.**

3. Add the Imagery basemap. Zoom to 1:2,000 scale.

4. Save your map document.

Deliverable 2: A map showing three types of wetland delineation. The map should have four data frames: TIN, Terrain, Kriging, and Comparison.

Draw conclusions and present the results

Once you complete your wetland comparison analysis, choose a method of presenting your conclusions. Always keep the audience in mind as you prepare to report your results; they may not share your GIS expertise. Your results can be presented in a Word document, a PowerPoint presentation, or a more technical presentation mode, such as ArcGIS Online.

DATA (Current Workspace)   \student\5_module10\1fl_data
RESULTS (Scratch Workspace)   \student\5_module10\1fl_results

MAKING
SPATIAL
DECISIONS
USING GIS
AND LIDAR

2

DEPRESSIONAL
WETLAND
DELINEATION
FROM LIDAR

PROJECT 2

Pasco County, Florida

The lidar data for this project is from Florida International University website: http://digir.fiu.edu/Lidar/LidarNew.php.

Wetlands data is from the US Fish and Wildlife Service website: http://www.fws.gov/wetlands/Data/Data-Download.html.

Recommended deliverables

- **Deliverable 1:** A basemap showing wetland types and LAS tiles.
- **Deliverable 2:** A map showing three types of wetland delineation. The map should have four data frames: TIN, Terrain, Kriging, and Comparison.

The questions asked in this project are both quantitative and qualitative. They identify key points that should be addressed in your analysis and final presentation.

Document your work, set environments, and examine the data

You need to add descriptive properties to every map document you produce. You can use the same descriptive properties for every map document in the project or customize the documentation from map to map.

For this project, **2fl_data** will be your project folder (EsriPress\MSDLidar\student \5_module10\2fl_data). Make sure that it is stored in a place where you have write access. You can store your output data inside the **2fl_results** folder (EsriPress\MSDLidar\student \5_module10\2fl_results). The results folder contains an empty geodatabase named **pasco_wetlands_results.gdb** in which to save your data.

DATA (Current Workspace)   \MSDLidar_Student\5_module10\2fl_data
RESULTS (Scratch Workspace)   \MSDLidar_Student\5_module10\2fl_results

181

MAKING
SPATIAL
DECISIONS
USING GIS
AND LIDAR

2

DEPRESSIONAL
WETLAND
DELINEATION
FROM LIDAR

1. Save your map documents inside the \student\5_module10\2fl_results folder. Name the map document **pasco1**.

2. Set environments:
   - Open the data frame properties. Set the map projection to Projected Coordinate Systems > State Plane > NAD 1983 (2011) (US Feet) > NAD_1983_2011_StatePlane_Florida_West_FIPS_0902_Ft_US.
   - On the Geoprocessing menu, click Environments.
   - Expand Workspace, and set the current workspace to \student\5_module10\2fl_data.
   - Set the scratch workspace to \student\5_module10\2fl_results.
   - For the output coordinate system, select Same as Display.

Analyze starting with a basemap

Construct a basemap

1. Connect to \student\5_module10\2fl_data.

In this folder, you will see another folder that holds the LAS files for Pasco County, Florida. The following is the metadata for the LAS files:

- Name: FL_Pasco_2004
- XY Coordinate System: NAD 1983 StatePlane Florida West FIPS 0902 (US Feet)
- Z Coordinate System: NAVD_1988 (US survey feet)
- Date of Acquisition: Load Data 2004
- Extent: Tile 4×4 square mile
- Filename: FL_Pasco_2004_000058.las

2. Add the following feature classes to the map document: pasco, wetlands, and pasco_lasd. Open the wetlands attribute table and look at the different types of wetlands.

Q1   *List the different types of wetlands.*

3. Save the map document as **pasco1** and then again as **pasco2**.

4. Before zooming to the study area, add the Imagery basemap.

Q2   *Describe Pasco County.*

DATA (Current Workspace)   \MSDLidar_Student\5_module10\2fl_data
RESULTS (Scratch Workspace)   \MSDLidar_Student\5_module10\2fl_results

MAKING
SPATIAL
DECISIONS
USING GIS
AND LIDAR

2

DEPRESSIONAL
WETLAND
DELINEATION
FROM LIDAR

Q3 ***What are the lowest and highest elevations shown in the two LAS datasets?***

Deliverable 1: A basemap showing wetland types and LAS tiles.

Derive TINs and contours from LAS datasets

1. Open the pasco2 map document and remove all layers except pasco.lasd.

2. Add the wetlands_sa feature class.

3. Construct a TIN using the following guidelines:
   - Set the LAS dataset filter to Ground.
   - Name the new TIN **pascoTIN**.

Frequently, the LAS dataset has too many points to save as a resulting TIN. You can fix this by performing a thinning process that reduces the LAS data points. When converting the pasco.lasd dataset to a TIN, use the following thinning parameters:
- Thinning type = RANDOM
- Thinning method = NODE_COUNT

Q4 ***What are the lowest and highest elevations shown in the pasco TIN?***

4. Create 1-foot surface contours. Select the contour closest to the NWI-designated wetland polygon (wetlands_sa feature class).

5. Convert the contour line to a polygon and calculate area in acres.

Derive contours from terrains

1. Insert a new data frame and name it **Terrain**.

2. Create a terrain with contour line, contour polygon, and calculate the area. Enter the following settings:
   - Set the point spacing to **4.9**.
   - Set the class code to **2**.

3. When creating the terrain, add 3 pyramids. (There should be 6 in total.)

DATA (Current Workspace) \MSDLidar_Student\5_module10\2fl_data
RESULTS (Scratch Workspace) \MSDLidar_Student\5_module10\2fl_results

183

MAKING
SPATIAL
DECISIONS
USING GIS
AND LIDAR

2

DEPRESSIONAL
WETLAND
DELINEATION
FROM LIDAR

Derive contours from kriging

1. Insert a new data frame and name it **Kriging**.

2. Add sa_kriging. Use it to clip the LAS to Multipoint tool output feature class containing points. Refer to project 1 for instructions on how to do this, if necessary.

3. Perform kriging, create contours, and calculate the area.

Compare wetland delineations

1. Insert a fourth data frame into your map document and name it **Wetland Comparison**.

2. Add the NWI wetland polygon, the TIN-derived polygon, the Terrain-derived polygon, and the Kriging-derived polygon.

Q5   ***Compare the contours visually. A good way to do this is to create a chart.***

Deliverable 2: A map showing three types of wetland delineation. The map should have four data frames: TIN, Terrain, Kriging, and Comparison.

Draw conclusions and present the results

After completing your wetlands delineation analysis for Pasco County, choose a method of presenting your conclusions. Always keep the audience in mind as you prepare to report your results; they may not share your GIS expertise. Your results can be presented in a Word document, a PowerPoint presentation, or a more technical presentation mode, such as ArcGIS Online.

DATA (Current Workspace)   \MSDLidar_Student\5_module10\2fl_data
RESULTS (Scratch Workspace)   \MSDLidar_Student\5_module10\2fl_results

MAKING
SPATIAL
DECISIONS
USING GIS
AND LIDAR

3

DEPRESSIONAL
WETLAND
DELINEATION
FROM LIDAR

PROJECT 3

On your own

Scenario

You have worked through a guided project and repeated the analysis for another project. In this project, you will reinforce your skills by researching and analyzing a similar scenario entirely on your own. First, identify your study area and acquire data for your analysis. You may want to study a local area.

Here is a list of topics that have been studied using lidar. The following represent possible project ideas:

- Forest characterization—canopy height and density
- Flood modeling
- Finding faults
- Geomorphic mapping
- Stream slope
- Archaeology field campaigns
- Mining—calculation of ore volumes
- Wind farm optimization

MAKING
SPATIAL
DECISIONS
USING GIS
AND LIDAR

3

DEPRESSIONAL
WETLAND
DELINEATION
FROM LIDAR

Many different websites distribute lidar data. Most of the available lidar data are within the United States. There are national data sites, and there are state data distributors. Reading the metadata and identifying a specific area of study is critical before downloading files. Some lidar files come compressed and require third-party software to convert them into the LAS format used by Esri.

- Open Topography facilitates community access to high-resolution, Earth science-oriented, topography data. It is a National Science Foundation-funded data facility. http://www.opentopography.org.

- The Earth Explorer provides online access to remotely sensed data from the US Geological Survey Earth Resources Observation and Science (EROS) Center archive. http://earthexplorer.usgs.gov/.

- National Oceanic and Atmospheric Administration, a world leader in coast science and management, provides state lidar datasets. http://www.csc.noaa.gov/dataviewer/#.

- Wikipedia lists national lidar datasets, organized according to state. http://en.wikipedia.org/wiki/National_LIDAR_Dataset_%E2%80%93_USA.

Research

Research the problem and answer the following questions:

1. What is the area of study?
2. What problem are you going to study?
3. What data is available?

IMAGE AND DATA CREDITS

Image credits

All images created by Keranen and Kolvoord.

Image (figure 03_basics_a.tif) from ArcGIS Help 10.2 http://resources.arcgis.com/en/help/main/10.2/index.html#/015w0000006n000000, courtesy of Esri.

Image (figure 02volume_shadow_a) from ArcGIS Help 10.2 http://resources.arcgis.com/en/help/main/10.2/index.html#/00q900000027000000, courtesy of Esri.

Image (figure 03volume_shadow_a) from ArcGIS Help 10.2 http://resources.arcgis.com/en/help/main/10.2/index.html#/How_Surface_Volume_works/00q900000037000000/, courtesy of Esri.

Image (figure 2_modules6_7_1a) from ArcGIS Help 10.2 http://resources.arcgis.com/en/help/main/10.2/index.html#/015w0000005q000000, courtesy of Esri.

Image (figure 05_graphic1) Scale 1:1,329, Virginia NAIP 2012 1m, Credits: One-meter resolution, leaf-on orthoimagery from 2012. Acquired by the USDA Farm Service agency Aerial Photography Field Office.

Image Service by VGIN.ADMIN.

Data credits

1_Modules_1-5 Baltimore

\EsriPress\MSDLidar\student\1_module1_5\0intro\documents\Lidar_Base_Specifications_Version_1.0, Heidemann, Hans Karl, 2012, Lidar base specification version 1.0: US Geological Survey Techniques and Methods, book 11, chap. B4, 63 p.

Basic lidar techniques, 2D and 3D models (raster and terrain), Volumetric analysis and shadow maps, Visibility analysis comparison, Surging seas

\EsriPress\MSDLidar\student\1_module1_5\baltimore_data\LAS_Files\MD_Baltimore_2008 _1S1E, courtesy of US Geological Survey.

\EsriPress\MSDLidar\student\1_module1_5\baltimore_data\LAS_Files\MD_Baltimore_2008 _1S1W, courtesy of US Geological Survey.

\EsriPress\MSDLidar\student\1_module1_5\baltimore_data\LAS_Files\MD_Baltimore_2008 _2S1E, courtesy of US Geological Survey.

\EsriPress\MSDLidar\student\1_module1_5\baltimore_data\LAS_Files\MD_Baltimore_2008 _2S1W, courtesy of US Geological Survey.

\EsriPress\MSDLidar\student\1_module1_5\baltimore_data\baltimore.gdb, Keranen and Kolvoord.

\EsriPress\MSDLidar\student\1_module1_5\baltimore_data\baltimore.gdb\baltimore_layers \baltimore_city, from Data and Maps for ArcGIS courtesy of Esri.

\EsriPress\MSDLidar\student\1_module1_5\baltimore_data\baltimore.gdb\baltimore_layers \bldgs9750, courtesy of City of Baltimore.

\EsriPress\MSDLidar\student\1_module1_5\baltimore_data\baltimore.gdb\baltimore_layers \bldgs9755, courtesy of City of Baltimore.

\EsriPress\MSDLidar\student\1_module1_5\baltimore_data\baltimore.gdb\baltimore_layers \bldgs, courtesy of City of Baltimore.

\EsriPress\MSDLidar\student\1_module1_5\baltimore_data\baltimore.gdb\baltimore_layers \bldgs1s1w, courtesy of City of Baltimore.

\EsriPress\MSDLidar\student\1_module1_5\baltimore_data\baltimore.gdb\baltimore_layers \highway, from Data and Maps for ArcGIS courtesy of Esri.

\EsriPress\MSDLidar\student\1_module1_5\baltimore_data\baltimore.gdb\baltimore_layers \sa_cell, Keranen and Kolvoord.

\EsriPress\MSDLidar\student\1_module1_5\baltimore_data\baltimore.gdb\baltimore_layers\towers, Keranen and Kolvoord.

\EsriPress\MSDLidar\student\1_module1_5\baltimore_data\baltimore.gdb\baltimore_layers\univ_plaz, Keranen and Kolvoord.

\EsriPress\MSDLidar\student\1_module1_5\baltimore_data\baltimore.gdb\baltimore_layers\univ_plaz_bldgs, courtesy of City of Baltimore.

MAKING
SPATIAL
DECISIONS
USING GIS
AND LIDAR

IMAGE AND
DATA CREDITS

\EsriPress\MSDLidar\student\1_module1_5\baltimore_data\baltimore.gdb\baltimore_layers\water, from Data and Maps for ArcGIS, courtesy of Esri.

\EsriPress\MSDLidar\student\1_module1_5\baltimore_data\baltimore.gdb\balt_elev_NED, courtesy of US Geological Survey.

\EsriPress\MSDLidar\student\1_module1_5\baltimore_data\baltimore.gdb\left_DEM, courtesy of US Geological Survey.

\EsriPress\MSDLidar\student\1_module1_5\baltimore_results\baltresults.gdb\, Keranen and Kolvoord.

1_Modules_1-5 San Francisco

Basic lidar techniques, 2D and 3D models (raster and terrain), Volumetric analysis and shadow maps, Visibility analysis comparison, Surging seas

\EsriPress\MSDLidar\student\1_module1_5\sanfrancisco_data\LAS_Files\ARRA-CA_SanFranCoast_2010_10SEG5282.las, courtesy of US Geological Survey.

\EsriPress\MSDLidar\student\1_module1_5\sanfrancisco_data\LAS_Files\ARRA-CA_SanFranCoast_2010_10SEG5283.las, courtesy of US Geological Survey.

\EsriPress\MSDLidar\student\1_module1_5\sanfrancisco_data\LAS_Files\ARRA-CA_SanFranCoast_2010_10SEG5382.las, courtesy of US Geological Survey.

\EsriPress\MSDLidar\student\1_module1_5\sanfrancisco_data\LAS_Files\ARRA-CA_SanFranCoast_2010_10SEG5383.las, courtesy of US Geological Survey.

\EsriPress\MSDLidar\student\1_module1_5\sanfrancisco_data\sanfrancisco.gdb, Keranen and Kolvoord.

\EsriPress\MSDLidar\student\1_module1_5\sanfrancisco_data\sanfrancisco.gdb\bldg100, Keranen and Kolvoord.

\EsriPress\MSDLidar\student\1_module1_5\sanfrancisco_data\sanfrancisco.gdb\bldg200, Keranen and Kolvoord.

\EsriPress\MSDLidar\student\1_module1_5\sanfrancisco_data\sanfrancisco.gdb\layers\bldgs, accessed from https://data.sfgov.org/Other/Data-Catalog/h4ui-ubbu.

\EsriPress\MSDLidar\student\1_module1_5\sanfrancisco_data\sanfrancisco.gdb\layers \bldgs_shadow, accessed from https://data.sfgov.org/Other/Data-Catalog/h4ui-ubbu.

\EsriPress\MSDLidar\student\1_module1_5\sanfrancisco_data\sanfrancisco.gdb \layers\counties, from Data and Maps for ArcGIS, courtesy of Esri.

\EsriPress\MSDLidar\student\1_module1_5\sanfrancisco_data\sanfrancisco.gdb \layers\gardens_2D, Keranen and Kolvoord.

\EsriPress\MSDLidar\student\1_module1_5\sanfrancisco_data\sanfrancisco.gdb\layers \highways, from Data and Maps for ArcGIS, courtesy of Esri.

\EsriPress\MSDLidar\student\1_module1_5\sanfrancisco_data\sanfrancisco.gdb\layers \sa_cell, Keranen and Kolvoord.

\EsriPress\MSDLidar\student\1_module1_5\sanfrancisco_data\sanfrancisco.gdb\sa_sanfrancisco, Keranen and Kolvoord.

\EsriPress\MSDLidar\student\1_module1_5\sanfrancisco_data\sanfrancisco.gdb\layers \san_francisco, from Data and Maps for ArcGIS, courtesy of Esri.

\EsriPress\MSDLidar\student\1_module1_5\sanfrancisco_data\sanfrancisco.gdb\layers \towers, Keranen and Kolvoord.

\EsriPress\MSDLidar\student\1_module1_5\sanfrancisco_data\sanfrancisco.gdb\layers \SF_elev_NED, courtesy of US Geological Survey.

2_Modules_6_7: JMU corrected 3D campus modeling and location of solar panels

\EsriPress\MSDLidar\student\2_module6_7\jmu_data\ARCHIVE_LAS_Files\ARCHIVE.lasd, Keranen and Kolvoord.

\EsriPress\MSDLidar\student\2_module6_7\jmu_data\ARCHIVE_LAS_Files\ LAS_N16_3874 _40.las, courtesy of William and Mary Center for Geospatial Analysis.

\EsriPress\MSDLidar\student\2_module6_7\jmu_data\JMU.gdb, Keranen and Kolvoord.

\EsriPress\MSDLidar\student\2_module6_7\jmu_data\JMU.gdb\layers\bldgs, courtesy of Virginia Information Technologies Agency.

\EsriPress\MSDLidar\student\2_module6_7\jmu_data\JMU.gdb\layers\dormitory, courtesy of Virginia Information Technologies Agency.

\EsriPress\MSDLidar\student\2_module6_7\jmu_data\JMU.gdb\layers\pond, courtesy of Virginia Information Technologies Agency.

\EsriPress\MSDLidar\student\2_module6_7\jmu_data\JMU.gdb\layers\study_area, Keranen and Kolvoord.

\EsriPress\MSDLidar\student\2_module6_7\jmu_data\JMU.gdb\layers\reclassify.lasd, Keranen and Kolvoord.

\EsriPress\MSDLidar\student\2_module6_7\jmu_results\JMU_results.gdb\layers, Keranen and Kolvoord.

MAKING
SPATIAL
DECISIONS
USING GIS
AND LIDAR

IMAGE AND
DATA CREDITS

2_Modules_6_7: SFU corrected 3D campus modeling and location of solar panels

\EsriPress\MSDLidar\student\2_module6_7\sfu_data\ARCHIVE_LAS_Files\archive.lasd, Keranen and Kolvoord.

\EsriPress\MSDLidar\student\2_module6_7\sfu_data\ARCHIVE_LAS_Files\ARRA-CA _GoldenGate_2010_000916.las, courtesy of US Geological Survey.

\EsriPress\MSDLidar\student\2_module6_7\sfu_data\las_to_classify\archive.lasd, Keranen and Kolvoord

\EsriPress\MSDLidar\student\2_module6_7\sfu_data\ARCHIVE_LAS_Files\ARRA-CA _GoldenGate_2010_000916.las, courtesy of US Geological Survey.

\EsriPress\MSDLidar\student\2_module6_7\sfu_data\SFU.gdb, Keranen and Kolvoord.

\EsriPress\MSDLidar\student\2_module6_7\sfu_data\SFU.gdb\layers\bldgs, courtesy of City of San Francisco.

\EsriPress\MSDLidar\student\2_module6_7\sfu_data\SFU.gdb\layers\sa_sanfrancisco, Keranen and Kolvoord.

\EsriPress\MSDLidar\student\2_module6_7\sfu_data\SFU.gdb\reclassifylas.lasd, Keranen and Kolvoord.

3_Modules8: New Jersey shoreline change after Hurricane Sandy

\EsriPress\MSDLidar\student\3_module8\nj_data\LAS_2010_PRE\20100828_2010_ncmp _nj_05_geoclassified_ld_p10.las, courtesy of Digital Coast NOAA Coastal Services Center.

\EsriPress\MSDLidar\student\3_module8\nj_data\LAS_2012_POST\20121120_2012_Post-Sandy_NJ_000278_GeoClassified.las, courtesy of Digital Coast NOAA Coastal Services Center.

\EsriPress\MSDLidar\student\3_module8\nj_data\NJ_Sandy.gdb, Keranen and Kolvoord.

\EsriPress\MSDLidar\student\3_module8\nj_data\NJ_Sandy.gdb\layers\monmouth, from Data and Maps for ArcGIS, courtesy of Esri.

\EsriPress\MSDLidar\student\3_module8\nj_data\NJ_Sandy.gdb\layers\NJ, from Data and Maps for ArcGIS, courtesy of Esri.

\EsriPress\MSDLidar\student\3_module8\nj_data\NJ_Sandy.gdb\layers\places, from Data and Maps for ArcGIS, courtesy of Esri.

\EsriPress\MSDLidar\student\3_module8\nj_data\NJ_Sandy.gdb\layers\study_area, Keranen and Kolvoord.

\EsriPress\MSDLidar\student\3_module8\nj_data\NJ_Sandy.gdb\layers\study_area2, Keranen and Kolvoord.

3_Modules8: New York shoreline change after Hurricane Sandy

\EsriPress\MSDLidar\student\3_module8\ ny_data\LAS_2010_PRE\20120120_18TXL015270. las, courtesy of Digital Coast NOAA Coastal Services Center.

\EsriPress\MSDLidar\student\3_module8\ny_data\LAS_2012_POST\20121120_2012 _PostSandy_NY_000097_GeoClassified.las, courtesy of Digital Coast NOAA Coastal Services Center.

\EsriPress\MSDLidar\student\3_module8\ny_data\NY.gdb, Keranen and Kolvoord.

\EsriPress\MSDLidar\student\3_module8\ny_data\NY.gdb\layers\pelham_manor, from Data and Maps for ArcGIS, courtesy of Esri.

\EsriPress\MSDLidar\student\3_module8\ny_data\NY.gdb\layers\study_area, Keranen and Kolvoord.

\EsriPress\MSDLidar\student\3_module8\ny_data\NY.gdb\layers\westchester, from Data and Maps for ArcGIS, courtesy of Esri.

\EsriPress\MSDLidar\student\3_module8\ny_data\NY_results.gdb\layers\westchester, from Data and Maps for ArcGIS, courtesy of Esri.

4_Modules9: Virginia forest vegetation height

\EsriPress\MSDLidar\student\4_Modules9\1va_data\LAS_files\LAS_N16_3803_10.las, courtesy of William and Mary Center for Geospatial Analysis.

\EsriPress\MSDLidar\student\4_Modules9\1va_data\va_forest.gdb, Keranen and Kolvoord.

\EsriPress\MSDLidar\student\4_Modules9\1va_data\va_forest.gdb\layers\GWNF, from Data and Maps for ArcGIS, courtesy of Esri.

\EsriPress\MSDLidar\student\4_Modules9\1va_data\va_forest.gdb\layers\studyarea_tiles, Keranen and Kolvoord.

\EsriPress\MSDLidar\student\4_Modules9\1va_data\va_forest.gdb\layers\va_counties, from Data and Maps for ArcGIS, courtesy of Esri.

MAKING
SPATIAL
DECISIONS
USING GIS
AND LIDAR

IMAGE AND
DATA CREDITS

\EsriPress\MSDLidar\student\4_Modules9\1va_data\va_forest.gdb\layers\va_countiesAnno, Keranen and Kolvoord.

\EsriPress\MSDLidar\student\4_Modules9\1va_data\va_forest.gdb\va_forest.lasd, Keranen and Kolvoord.

\EsriPress\MSDLidar\student\4_Modules9\1va_results\va_forest.gdb\va_forest_results.gdb, Keranen and Kolvoord.

MAKING
SPATIAL
DECISIONS
USING GIS
AND LIDAR

IMAGE AND
DATA CREDITS

4_Modules9: Pennsylvania forest vegetation height

\EsriPress\MSDLidar\student\4_Modules9\2pa_data\LAS_files\24002060PAS.las, courtesy of The Pennsylvania Geospatial Data Clearinghouse.

\EsriPress\MSDLidar\student\4_Modules9\2pa_data\LAS_files\24002070PAS.las, courtesy of The Pennsylvania Geospatial Data Clearinghouse.

\EsriPress\MSDLidar\student\4_Modules9\2pa_data\pa_forest.gdb, Keranen and Kolvoord.

\EsriPress\MSDLidar\student\4_Modules9\2pa_data\pa_forest.gdb\layers\michaux_sf, courtesy of The Pennsylvania Geospatial Data Clearinghouse.

\EsriPress\MSDLidar\student\4_Modules9\2pa_data\pa_forest.gdb\layers\pa_breaklines, courtesy of The Pennsylvania Geospatial Data Clearinghouse.

\EsriPress\MSDLidar\student\4_Modules9\2pa_data\pa_forest.gdb\layers\pa_counties, from Data and Maps for ArcGIS, courtesy of Esri.

\EsriPress\MSDLidar\student\4_Modules9\2pa_data\pa_forest.gdb\layers\places, from Data and Maps for ArcGIS, courtesy of Esri.

\EsriPress\MSDLidar\student\4_Modules9\2pa_data\pa_forest.lasd, Keranen and Kolvoord.

\EsriPress\MSDLidar\student\4_Modules9\2pa_results\pa_forest_results.gdb, Keranen and Kolvoord.

5_Modules10: Florida 1 depressional wetland delination from lidar

\EsriPress\MSDLidar\student\5_modules10\1fl_data\LAS_files\LID2007_065029_N.las, courtesy of Florida International University GIS Center.

\EsriPress\MSDLidar\student\5_Modules10\1fl_data\fl_wetlands.gdb, Keranen and Kolvoord.

\EsriPress\MSDLidar\student\5_Modules10\1fl_data\fl_wetlands.gdb\layers\counties, from Data and Maps for ArcGIS, courtesy of Esri.

\EsriPress\MSDLidar\student\5_Modules10\1fl_data\fl_wetlands.gdb\layers\sa_kriging, Keranen and Kolvoord.

\EsriPress\MSDLidar\student\5_Modules10\1fl_data\fl_wetlands.gdb\layers\study_area, Keranen and Kolvoord.

\EsriPress\MSDLidar\student\5_Modules10\1fl_data\fl_wetlands.gdb\layers\tile29_wetlands _sa, courtesy of US Fish & Wildlife Service National Wetlands Inventory.

\EsriPress\MSDLidar\student\5_Modules10\1fl_data\fl_wetlands.gdb\layers\wakulla, from Data and Maps for ArcGIS, courtesy of Esri.

\EsriPress\MSDLidar\student\5_Modules10\1fl_data\fl_wetlands.gdb\layers\wetlands, courtesy of US Fish & Wildlife Service National Wetlands Inventory.

\EsriPress\MSDLidar\student\5_Modules10\1fl_data\tile_29.lasd, Keranen and Kolvoord.

\EsriPress\MSDLidar\student\5_Modules10\1fl_result\fl_wetlands_results.gdb, Keranen and Kolvoord.

5_Modules10: Florida 2 depressional wetland delination from lidar

\EsriPress\MSDLidar\student\5_modules10\2fl_data\LAS_files\FL_Pasco_2004_000058.las, courtesy of Florida International University GIS Center.

\EsriPress\MSDLidar\student\5_Modules10\2fl_data\fl_wetlands.gdb, Keranen and Kolvoord.

MAKING
SPATIAL
DECISIONS
USING GIS
AND LIDAR

*IMAGE AND
DATA CREDITS*

\EsriPress\MSDLidar\student\5_Modules10\2fl_data\fl_wetlands.gdb\layers, Keranen and Kolvoord.

\EsriPress\MSDLidar\student\5_Modules10\2fl_data\fl_wetlands.gdb\Layers\counties, from Data and Maps for ArcGIS, courtesy of Esri.

\EsriPress\MSDLidar\student\5_Modules10\2fl_data\fl_wetlands.gdb\Layers\pasco, from Data and Maps for ArcGIS, courtesy of Esri.

\EsriPress\MSDLidar\student\5_Modules10\2fl_data\fl_wetlands.gdb\Layers\sa_kriging, Keranen and Kolvoord.

\EsriPress\MSDLidar\student\5_Modules10\2fl_data\fl_wetlands.gdb\Layers\wetlands, courtesy of US Fish & Wildlife Service National Wetlands Inventory.

\EsriPress\MSDLidar\student\5_Modules10\2fl_data\fl_wetlands.gdb\Layers\wetlands_sa, courtesy of US Fish & Wildlife Service National Wetlands Inventory.

\EsriPress\MSDLidar\student\5_Modules10\2fl_data\pasco.lasd, Keranen and Kolvoord.

\EsriPress\MSDLidar\5_Modules10\2fl_results\pasco_wetlands_results.gdb, Keranen and Kolvoord.

DATA LICENSE AGREEMENT

Important: Read carefully before downloading the media.

Environmental Systems Research Institute, Inc. (Esri), is willing to license the enclosed data and related materials to you only upon the condition that you accept all of the terms and conditions contained in this license agreement. Please read the terms and conditions carefully before downloading the media. By downloading the media, you are indicating your acceptance of the Esri License Agreement. If you do not agree to the terms and conditions as stated, then Esri is unwilling to license the data and related materials to you.

Esri license agreement

This is a license agreement, and not an agreement for sale, between you (Licensee) and Environmental Systems Research Institute, Inc. (Esri). This Esri License Agreement (Agreement) gives Licensee certain limited rights to use the data and related materials (Data and Related Materials). All rights not specifically granted in this Agreement are reserved to Esri and its Licensors.

Reservation of ownership and grant of license

Esri and its Licensors retain exclusive rights, title, and ownership to the copy of the Data and Related Materials licensed under this Agreement and, hereby, grant to Licensee a personal, nonexclusive, nontransferable, royalty-free, worldwide license to use the Data and Related Materials based on the terms and conditions of this Agreement. Licensee agrees to use reasonable effort to protect the Data and Related Materials from unauthorized use, reproduction, distribution, or publication.

Proprietary rights and copyright

Licensee acknowledges that the Data and Related Materials are proprietary and confidential property of Esri and its Licensors and are protected by United States copyright laws and applicable international copyright treaties and/or conventions.

Permitted uses

Licensee may install the Data and Related Materials onto permanent storage device(s) for Licensee's own internal use.

Licensee may internally use the Data and Related Materials provided by Esri for the stated purpose of GIS training and education.

Uses not permitted

Licensee shall not sell, rent, lease, sublicense, lend, assign, time-share, or transfer, in whole or in part, or provide unlicensed Third Parties access to the Data and Related Materials or portions of the Data and Related Materials, any updates, or Licensee's rights under this Agreement.

Licensee shall not remove or obscure any copyright or trademark notices of Esri or its Licensors.

Term and termination

The license granted to Licensee by this Agreement shall commence upon the acceptance of this Agreement and shall continue until such time that Licensee elects in writing to discontinue use of the Data or Related Materials and terminates this Agreement. The Agreement shall automatically terminate without notice if Licensee fails to comply with any provision of this Agreement. Licensee shall then return to Esri the Data and Related Materials. The parties hereby agree that all provisions that operate to protect the rights of Esri and its Licensors shall remain in force should breach occur.

Disclaimer of warranty

The Data and Related Materials contained herein are provided "as-is," without warranty of any kind, either express or implied, including, but not limited to, the implied warranties of merchantability, fitness for a particular purpose, or noninfringement. Esri does not warrant that the Data and Related Materials will meet Licensee's needs or expectations, that the use of the Data and Related Materials will be uninterrupted, or that all nonconformities, defects, or errors can or will be corrected. Esri is not inviting reliance on the Data or Related Materials for commercial planning or analysis purposes, and Licensee should always check actual data.

Data disclaimer

The Data used herein has been derived from actual spatial or tabular information. In some cases, Esri has manipulated and applied certain assumptions, analyses, and opinions to the Data solely for educational training purposes. Assumptions, analyses, opinions applied, and actual outcomes may vary. Again, Esri is not inviting reliance on this Data, and the Licensee should always verify actual Data and exercise their own professional judgment when interpreting any outcomes.

Limitation of liability

Esri shall not be liable for direct, indirect, special, incidental, or consequential damages related to Licensee's use of the Data and Related Materials, even if Esri is advised of the possibility of such damage.

No implied waivers

No failure or delay by Esri or its Licensors in enforcing any right or remedy under this Agreement shall be construed as a waiver of any future or other exercise of such right or remedy by Esri or its Licensors.

Order for precedence

Any conflict between the terms of this Agreement and any FAR, DFAR, purchase order, or other terms shall be resolved in favor of the terms expressed in this Agreement, subject to the government's minimum rights unless agreed otherwise.

Export regulation

Licensee acknowledges that this Agreement and the performance thereof are subject to compliance with any and all applicable United States laws, regulations, or orders relating to the export of data thereto. Licensee agrees to comply with all laws, regulations, and orders of the United States in regard to any export of such technical data.

MAKING
SPATIAL
DECISIONS
USING GIS
AND LIDAR

DATA LICENSE
AGREEMENT

Severability

If any provision(s) of this Agreement shall be held to be invalid, illegal, or unenforceable by a court or other tribunal of competent jurisdiction, the validity, legality, and enforceability of the remaining provisions shall not in any way be affected or impaired thereby.

Governing law

This Agreement, entered into in the County of San Bernardino, shall be construed and enforced in accordance with and be governed by the laws of the United States of America and the State of California without reference to conflict of laws principles. The parties hereby consent to the personal jurisdiction of the courts of this county and waive their rights to change venue.

Entire agreement

The parties agree that this Agreement constitutes the sole and entire agreement of the parties as to the matter set forth herein and supersedes any previous agreements, understandings, and arrangements between the parties relating hereto.

MAKING
SPATIAL
DECISIONS
USING GIS
AND LIDAR

DATA LICENSE
AGREEMENT